Oedipus in
Evolution

Oedipus in Evolution

A New Theory of Sex

Christopher Badcock

First published 1990

Basil Blackwell Ltd
108 Cowley Road, Oxford OX4 1JF, UK

Basil Blackwell Inc.
3 Cambridge Center,
Cambridge, Massachusetts 02142, USA

British Library Cataloguing in Publication Data

A CIP catalogue record for this book is available
from the British Library.

Library of Congress Cataloging in Publication Data

Badcock, C. R.
  Oedipus in Evolution : a new theory of sex / Christopher Badcock.
      p.   cm.
  Bibliography: p.
  Includes index.
  ISBN 0–631–15794–8
  1. Sex (psychology)   2. Sociobiology   3. Psychoanalysis.
4. Freud, Sigmund, 1856–1939.   I. Title.
  BF692.B29   1989
  155.3–dc20   89-15032
                                                            CIP

Typeset by Ponting–Green Publishing Services, London, England
Printed in Great Britain by Billing and Sons Ltd., Worcester

Contents

Acknowledgements

Brief summaries of some of the more controversial ideas found here were given at the Thirty-second Annual Meeting of The American Academy of Psychoanalysis in Montreal and to members of the Evolution and Human Behavior Program at the University of Michigan during May 1988. I must thank Arthur Epstein, Myron Glucksman and Saul Tuttman for their kind hospitality in inviting me to the Montreal meeting and for financial assistance in enabling me to attend it. I am deeply indebted to Randolph Nesse, Director of the University of Michigan Evolution and Human Behavior Program, who was instrumental, both in inviting me to the Montreal conference and to Ann Arbor, and in providing considerable financial assistance for my visit. I owe him a further special debt of gratitude for discussing his ideas on oral behaviour with me and for introducing me to the other participants in the evolution and psychoanalysis panel at Montreal: Alan Lloyd and Malcolm Slavin, to both of whom I am alsó much indebted.

Early drafts of the book were read by my wife and by Tracey Fox, Ben Hoffschulte, Alan Lloyd, Randolph Nesse, Daniel Rancour-Laferriere, Keith Sharp and Jan Wind. I am indebted to them, as well as to the participants of the Montreal conference and to members of the Evolution and Human Behavior Program at Ann Arbor for their helpful suggestions, criticisms and observations. I must thank Thomas Gregor, Valerius Geist, Warren Shapiro and Robert Trivers for their kind permission to quote from their works and Roger V. Short for permission to use material on which the figure on page 57 is based and, along with Shyam Thapa and Malcolm Potts, for permission to quote from their recent paper.

Finally, I must thank my publisher, John Davey, for his unstinting support, help and guidance throughout.

Christopher Badcock
February 1989

'Psychology will be based on a new foundation ...'
Charles Darwin, *The Origin of Species*

Introduction:
Evolution in Psychoanalysis

The last three minutes

As far as human history is concerned, everything appears to have happened in the last few thousand years – certainly within the ten thousand-odd years since the Neolithic Revolution brought with it the records, monuments and artifacts without which recorded history would not exist. Yet, from the point of view of natural history, our tenure runs to at least a million years – two orders of magnitude more. So exiguous is the period of recorded history that, in giving a series of ten one-hour lectures on human societies, I used to point out to my audience that if I allocated time strictly according to the probable duration of forms of social structure then all post-Neolithic societies together would get approximately the last three minutes of the last lecture and the primal hunter–gatherer societies which preceded them the remaining nine hours and fifty-seven minutes!

Where recorded history is concerned, it seems that nothing happens for a million years or so and then everything happens very quickly in a few thousand. Indeed, not only does everything happen very suddenly in comparatively recent times, but the pace and intensity of historical events, represented by the numbers of human beings living, their life-expectancy and the complexity of their interactions with one another and with their environment, increase almost exponentially. The consequence is that now, on the eve of the third millennium, more human beings are alive than at any one time in the past; and more of them, via modern communications and transport, come into contact with vastly more others than has ever been possible previously.

In this respect the social universe seems the exact opposite of the physical cosmos. In a book entitled *The First Three Minutes* the physicist Steven Weinberg has portrayed the events which are now believed to have marked the origin of time, space and everything we know. It seems a startling paradox that, by contrast to human evolution, in the evolution of the universe as a whole the vast majority of events (understood as interactions between the primary constituents of matter and energy) were over by the end of the first few minutes! So great were the density, temperature and energy of the universe at that time that its present average temperature of a mere three degrees above absolute zero makes it seem a cold, dark and empty place by comparison with its hot, dense and violent beginning.

Yet, just as those first few moments of cosmic time were critical for all that was to follow and seem to have determined the basic physical constraints within which the universe has worked ever since (such as the preponderance of matter over anti-matter), so it seems that the first million years or so of human existence have come to exercise a fundamental and determining influence over the relatively brief, but dynamic, period of recorded history. In complete contrast to physical history, where the first and most fundamental events were also the briefest, evolutionary history is characterized by slow and gradual change, at least by comparison with the hectic pace of recorded history or the even more astonishing pace of events immediately after the creation of the cosmos.

In biological evolution, the final arbiter is a selective force whose influence seems slow and gradual from a human point of view because it only shows itself in the effect of *differential reproductive success* (in other words, in the realization that some organisms leave more descendants than others). Since such differential effects can only manifest themselves on a time-scale determined by the generational turn-over of the organisms in question, the period involved must tend to become a very long one if, as in the human case, the generational time-span is measured in decades.

Charles Darwin's basic insight was that, since organisms can reproduce faster or in greater numbers than the environment can usually sustain, those organisms which show some kind of adaptive advantage in the given circumstances will prosper at the expense of the rest. As a consequence, the circumstances in question – and they are multifarious, and in principle can take more or less any form – come to exercise a selective function, rather as human breeders of animals do in choosing certain characteristics in the individuals they manipulate. The result is the vast profusion of different organisms which we see around us today, each owing its distinct identity as a species to particular selective circumstances which have shaped the reproductive success of its ancestors and have given their descendants their particular form.

It follows from these considerations that, since the pace of evolutionary change is so slow by comparison with the tempo of recorded history, the last twelve to fifteen thousand years can have had little fundamental effect on the genetic adaptations of human beings. The four or five hundred generations which at the most separate us from our primeval hunter–gatherer ancestors do not constitute sufficient time for more than relatively minor evolutionary changes, such as those involved in skin colour, to occur. Furthermore, those human populations whose primal hunter–gatherer ancestors can be found only a few generations back – the aborigines of Australia and possibly one or two other groups – do not show any real sign of being fundamentally different from all other human beings as far as their basic adaptations are concerned.

It seems, then, that if we wish to undertake an exercise analogous to that of modern cosmologists in tracing back the current human universe to its moment of creation, that 'moment' – in stark contrast to the physical universe – is in fact some millions of years long, whereas the period of rapid modern expansion of human numbers coincident with recorded history is astonishingly short. The existence of a possible two million or so years of hunter–gatherer prehistory compared to the ten-thousand-odd years of post-hunter–gatherer time means that those who want to understand our species' origins need to redress the imbalance of recorded history and direct their attention instead to that primal epoch when humanity as we know it came into being. In the following pages this bias will be followed and, analogously with the modern cosmological concern with the first three minutes of existence, this book will mainly concern itself with what must have come about before the period of recorded history, itself merely the last few moments of evolutionary time.

The fourth dimension of psychoanalysis

In pursuing a psychological, as opposed to a purely physical, investigation of mental disorders, Freud found that each of his patients had a history which, when presented in a written form, seemed to recall the novel or short story more than the traditional medical case-history. Right up to the present this has remained a cause of scandal to those who hold that mental disorders can only originate in physical or genetic pathology. To those who take a wider view, however, the psychological case-history has come to occupy its proper, central place in the diagnosis and explanation of many kinds of psychopathology.

Yet, if the arguments outlined above are correct, there is another kind of history which ought also to be taken into account. Freud's patients had not only a personal history, they had, collectively, a *natural history*. They were not merely Viennese men, women and – in one case at least – children. They were also human beings: that is, members of a species which evolved its particular adaptations for a way of life strikingly different from that which its members were experiencing in late nineteenth- and early twentieth-century Vienna.

In describing and explaining the psychology of his patients Freud gradually evolved three principal means: the so-called *dynamic, topographical* (or *structural*) and *quantitative* (or *economic*) dimensions of psychoanalytic theory. Of these, the dynamic was in many ways the most significant and distinctive of psychoanalysis. It was a means of description and explanation which saw the mind as riven by the active and reactive interplay of conflicting forces. Freud's early work on hysteria, for instance, convinced him that hysterical symptoms – and possibly, the symptoms of most forms of mental illness – were the outcome of conflicting currents of feeling and thinking which resulted in certain elements being excluded from consciousness and acquiring powerful instinctual reinforcement in the unconscious. In the dynamic dimension of explanation repression plays a central role, but many other forces and processes are at work, such as reaction-formation, projection and identification.

The consequence of psychological dynamics is mental structure, or *topography*: that is, the tendency for certain elements of consciousness to be driven into the unconscious by countervailing forces and for unconscious factors to affect conscious ones by indirect means. The topographic dimension of psychoanalytic psychology sees the mind as structured in terms of spatial, rather than dynamic interrelations. If the dynamic dimension gives rise to the concept of mental *forces*, then topography correspondingly suggests the idea of *fields*. It pictures the mind as stratified into different regions of consciousness or divided up among differing systems in a way which can (in principle at least) be mapped.

Finally, Freud had recourse to the quantitative or economic dimension of explanation: the view which sees the mind as containing definite quantities or charges of excitation whose values were significant both for dynamic interactions (for instance, in deciding which of two forces is the stronger) and for the topographical dimension in setting exact limits to the fields of consciousness and unconsciousness, instinct and ego and so on. If dynamics means force and topography field, then the quantitative view suggests the analogy of the *vector* – that is, something with both a direction and a quantity.[1]

Yet, from another point of view, the three dimensions of psychoanalytic explanation can be seen as analogous to the three dimensions of ordinary space around us. They do not take any account of time, at least not in the evolutionary sense. Rather like physics before Einstein's general theory of relativity blended a fourth, temporal dimension with the three conventional spatial ones, psychoanalysis until now has ignored the immense span of evolutionary time during which mental dynamics, topography and economics evolved into their present form.

In the case of physics, the integration of time with the spatial dimensions followed because it became necessary to recognize that measurements of both size and duration were subject to interrelated variation – time was 'mixed up' with space, so to speak. The resulting four-dimensional 'curvature' was far too subtle to be seen on anything except a cosmic scale, and hard enough to detect even there. Yet, for all that, the integration of the fourth, temporal dimension as if it were another spatial one was mathematically unavoidable and theoretically quite justified. In a corresponding way the foregoing discussion of human evolutionary time suggests that to the three 'spatial' and conventional dimensions of psychoanalytic theory we should add a fourth, temporal one and perhaps call it the *evolutionary* dimension.

[1] For a short account of Freud's fundamental theory see C. Badcock, *Essential Freud*, chapter 1.

In this context it is worth pointing out that, although the development of this new, fourth perspective had to wait until progress in both evolutionary theory and our knowledge of human prehistory had become much more reliable and extensive than during Freud's lifetime, he had glimpsed its possibilities as early as 1915. In that year he had drafted a series of theoretical papers devoted to so-called *metapsychology* – that is, the explanation of the observed facts of human psychology in dynamic, topographical and quantitative terms. Few of these ever saw the light of day, partly because of wartime conditions, and also the rapid progress in theoretical psychoanalysis during and just after the war. However, a draft of the final paper was recently discovered which, entitled 'Overview of the Transference Neuroses', was obviously intended to be a summary and conclusion to the whole series.[2]

This paper begins in conventional metapsychological terms with considerations relating to quantitative factors like libidinal charge and dynamic ones like regression and repression, and assumes the so-called first psychoanalytic topography throughout. But then it suddenly switches to what I have termed the fourth, evolutionary view and attempts to reconstruct the adaptive history underlying the problem which the paper attempts to resolve – that of the pattern of onset and development of human psychopathology. Although the substantive answers which Freud attempted to give to this difficult problem may seem quite naive to modern eyes (and certainly could not have impressed their author for long, because he never attempted to publish them), in terms of their form rather than their content they seem to be many decades ahead of their time and essentially prefigure the line of reasoning being developed here.

The new evolutionary dimension sees mental phenomena in terms not of forces, fields or vectors, but as classical Darwinian *adaptations*: that is, mental processes which have evolved in order to enhance the reproductive success of their possessors and which have become the common evolutionary heritage of all modern human beings. It sees them against the background not of personal, but of evolutionary history, and adds a hitherto neglected and unnoticed dimension to our understanding of ourselves as the products of evolution. It leads to a new insight into our mental life which sees it as not

[2] S. Freud, *A Phylogenetic Fantasy.*

merely the outcome of conflicting forces, differentiated fields and variable quantities of excitation, but as the end product of a lengthy process of adaptation to a hunter–gatherer prehistory.

Such an evolutionary approach can answer questions never before answerable in any but the most preliminary and general form. Questions concerning the ultimate determinants of fundamental psychological processes such as identification and repression, or the causes of Oedipal behaviour, penis-envy, homosexuality and other deviations have only been answerable in dynamic, topographical and economic terms which had to take more fundamental assumptions for granted. The assumptions I have in mind are things like constitutional human bisexuality and the dynamic, topographical and economic factors themselves.

Until recently, no one could really say *why* human beings have Oedipus complexes, except in the most general way and in terms of such basic assumptions as those which I have just mentioned. In the past we could not really say why repression, identification or sexual deviations operate in the way in which they do, because most of our knowledge in psychoanalysis has been in terms of *how* such things occur. Again, we think we know how the work of mourning is accomplished and why it takes so long, but we have never really known why it had to exist in the first place or why time seemed so important to it.

One major benefit of an evolutionary view is that it can answer these kinds of questions in something like conclusive terms. It can indeed tell us why – and possibly even how – Oedipal behaviour, identification or repression evolved, and what their adaptive benefits were. It can tell us why human beings are bisexual in Freud's understanding of that term and why mourning and the depression which is a major part of it have a periodic dimension and what its adaptive benefits are. Finally, it can tell us why mental dynamics, psychological topography and economics are disposed in our species as they are by virtue of seeing all these things as adaptations which evolved to promote the evolutionary success of our ancestors. Their adaptive meaning was not immediately clear in late nineteenth- and early twentieth-century Vienna simply because the conditions in which human psychology emerged were rather different to those of Freud's Vienna and were only to be found in the primal hunting and gathering societies in which our ancestors existed for the vast majority of the time which our species has spent on the Earth.

Once we understand that the ultimate parameters of human psycho-logy and behaviour have been set for such conditions our insight into human nature can be immeasurably deepened and extended. It seems that, just as the early successes of psychoanalysis were won by delving into the personal history of the individual, so the ultimate success of its insights might be achieved by exploring the natural history of our species by means of a fourth, final and fundamental dimension of psychoanalysis – the evolutionary one suggested here.[3]

The re-interpretation of dreams

In the famous seventh and last chapter of *The Interpretation of Dreams* Freud set out what is usually called his first topography: a model of the mind which represents it as structured on the basis of strati-graphic distinctions between conscious–perceptual, preconscious and unconscious systems. Later, around 1920, he introduced his so-called second topography, the well-known id–ego–superego model. Recently, a number of sociobiologists have commented on the aptness of these topographies to modern biological understandings of the likely evolutionary forces shaping the development of conscious-ness and, in particular, the adaptive advantages of the unconscious.[4]

In his quantitative, as opposed to topographical, models Freud made use of an idea which was borrowed directly from biology: the concept of the instinctual drive. These drives arose in what Freud termed the *id* and addressed themselves to the *ego*. In my earlier study which complements this one,[5] I proposed some reasons why a slight reformulation of these concepts might be helpful. I proposed what may eventually come to be seen as a third topography based on what I would term the EGO-centric approach. Here the acronym EGO stands for a rather more carefully defined and restricted version of what, in the second topography, Freud meant by ego, namely, an *Executive and Governing Organization*.

Like Freud's ego, which it intentionally suggests and which was conceived by him essentially as an organization,[6] the new EGO is a

---

[3] Badcock, *Essential Freud*, pp. 166–77.

[4] See, for instance, R. Trivers, *Social Evolution*, pp. 163–4, and D. Barash, *Sociobiology and Behavior* (2nd edn) p. 157.

[5] Badcock, *The Problem of Altruism*, pp. 164–71.

[6] Freud, *Inhibitions, Symptoms and Anxiety*, XX, 97. Unless otherwise stated, this and all subsequent references to the works of Freud are to *The Standard Edition*

mental agency whose nature is defined by its function: that of being at the interface of external perception and internal sensation and in control of voluntary movement and volitional thought. Its motto is not so much the purely intellectual *Cogito ergo sum* ('I think, therefore I am'), but the more active *Ego ergo ago* ('I am, therefore I act'). It is emphatically not synonymous with the 'ego' of popular expression – the personification of the the self (for which I have suggested the alternative term 'persona') – and is largely unconscious. Like Freud's ego, it is purely psychological and exclusively a model of mental function, rather than some neurological structure.

First principles of evolutionary biology suggest that such a mental organization as the human EGO could not emerge accidentally and without adaptive value. The proof of this is the realization that, were the EGO to be non-adaptive, the resources tied up in it would place its possessors at a disadvantage with regard to others who lacked such a psychological ornament. Since the costs of EGO-functioning seem to be very high, because of the size of brain needed to sustain it and the extent of its interference in behaviour, we may assume that such an organization of mental functioning could not evolve by accident and could not be adaptively neutral.

Assuming then that the EGO is adaptive and having defined it as an executive and governing organization of behaviour, we must now ask how evolution could have influenced its activity. We now encounter the problem of instinct, and here a natural and logical conclusion follows if we adhere to an approach which looks at things from the point of view of the EGO. So rather than become entangled in the controversies over instincts and drives I suggest that we merely assume that the EGO has certain constraints placed upon it by the organism, constraints which are designed, in their evolutionary origin at least, to maximize the reproductive success of the organism's genes.

For the time being, it does not matter in the least what these constraints and demands actually are; all that is relevant is the realization that they must tend on average, and in the context of primal human existence, to discourage behaviour which puts the organism and its genes at a disadvantage in evolutionary terms and

of the Complete Psychological Works of Sigmund Freud, quoting title, volume and page number.

to encourage behaviour which does the opposite.

Reverting to our earlier mention of the principle of evolution, we can easily prove the credibility of this assumption. As I remarked earlier, it is reproductive success which drives evolution, under the influence of selective factors. Evolutionary reproductive success is often called 'Darwinian fitness', and the idea of *inclusive* fitness extends this to include the reproductive success of copies of genes – the fundamental elements reproduced – present in near relatives of the individual in question. In other words, the concept of inclusive fitness reflects not merely the fact that natural selection selects for reproductive success, but that the genes selected are shared by closely related individuals who carry other copies of them. If we conceive of selection operating at the level of the gene, then a gene selected in one individual can be selected in another who has an identical copy of it.

To sum up, we can say that any behaviour which promotes inclusive fitness will tend to be selected and any which reduces it will tend to be selected against. In other words, given the choice of demands placed on the EGO to promote or reduce the individual's reproductive success, only the former will be favoured by evolution. We can conclude that, having defined the EGO in purely neutral terms as far as the fitness-maximizing nature of its behaviour is concerned (a definition, I might add, which exactly fits our subjective awareness), we are forced to conclude that such a structure could not be expected to evolve without a complementary system of adaptive demands placed upon it. These constraints I propose to call *Inclusive-fitness-maximizing Demands* or the ID for short – this ID standing in relation to Freud's id as our new Executive and Governing Organisation does to his ego.

Alternatively, ID might be seen as standing for *Instinctual Demand* in the sense in which we find Freud observing that 'an instinct is without quality, and, so far as mental life is concerned, is only to be regarded as a measure of the demand made upon the mind for work.'[7] The 'work' in question has only to be interpreted as anything which will meet the demands of the maximization of inclusive fitness for this concept of instinct to be entirely in accordance with modern biological understanding of the evolution of behaviour.

The fact that evolution is driven by the differential reproductive

[7] Freud, *Three Essays on the Theory of Sexuality*, VII, 6

success of organisms competing for scarce resources means that an organism must concern itself, not merely with its own existence, but with reproduction. Looked at from this point of view, both the *libido theory* and the *pleasure principle* seem obvious. Because evolution is dependent on competitive reproductive success, organisms must be motivated to reproduce. This implies both a sexual demand upon the EGO and some means of motivating it to meet such demands. In safeguarding individual existence pain and unpleasure in general seem to be prime motivating factors. The EGO recoils from conditions which cause it to perceive pain, unpleasurable tension, stress and anxiety.

But such negative motivations apply mainly to avoiding life-endangering or self-damaging circumstances; the avoidance of unpleasure does not seem to be a main factor in sexual motivation, at least in the first instance. Here a positive motivation exists in pleasure, which seems to be strongly implicated in human sexual behaviour so that the EGO is rewarded with pleasurable sensations whenever it satisfies the fitness-maximizing sexual demands of the ID. This must be the essential evolutionary basis of the pleasure principle: it serves both to mobilize fitness-maximizing behaviour with pleasurable rewards offered to the EGO and to discourage fitness-minimizing behaviour with painful, anxiety-provoking and unpleasurable stresses directed towards it.

For Freud, with his metapsychological assumptions relating to the ego (as he conceived it), tension was a problem because he assumed that the organism was adapted to desire the least possible excitation – the so-called 'Nirvana principle'. He viewed the ego as something with a built-in tendency to reduce tension to the lowest possible value. This was probably because he was influenced in his early years by biological and psychological theories which sought to see the mind as some kind of natural mechanism which obeyed basic physical principles. One of these evidently is that a system can be expected to revert to its lowest energy level in the absence of external inputs, simply because energy always has to come from somewhere and has to be paid for. Since tension represents a cost to the ego it is logical to suppose that, like a loaded spring which is suddenly released, the mind would revert to its lowest, equilibrium state once instinctual demands on it were met.

Unfortunately, and expressed in these perhaps too simple terms, the Nirvana principle seems to predict absurdities. For instance, since death is the lowest energy state available to a living system, death would appear to be preferable to life (a consideration which seems to have influenced Freud in his speculations regarding a 'death instinct'). Again, the principle contradicts observed fact because we notice that, in the sexual sphere for instance, excitation and tension are inherently pleasurable, sometimes irrespective of eventual satisfaction via reduction in tension (orgasm). Indeed, if pleasure were not pleasurable, no one would want to continue it; yet the fact is that pleasure does make us want to continue it, even if it keeps us in a state of tension.

The substitution of the concept of inclusive-fitness-maximizing demands for 'drives' or 'instincts' as understood by Freud and the biology of his day immediately removes these problems because it pictures the new EGO as experiencing unpleasurable tension in relation to the demand, not to its execution. For instance, we assume that the ID addresses fitness-maximizing demands to the EGO for sexual activity but, once the EGO begins to meet the demand, the tension involved in doing so becomes inherently pleasurable.

To understand this we have to notice one important peculiarity about the ID. Whereas it appears to have numerous 'up-link' channels by means of which it can address demands to the EGO (for instance, via reflexes like salivation and erection, via fantasies, emotional states and so on), the EGO seems to lack comparable 'down-link' channels by means of which it can modify the ID. On the contrary, the essential finding of psychoanalysis is that the only means by which the EGO can modify the ID are by repression and defence. In other words, it can cut, modify or divert the ID's 'up-link' channels, as I am calling them, but it cannot re-program the ID as such by means of a 'down-link' of its own.

The consequence of this is that it may well be that the tension involved in actually acting on a fitness-maximizing demand is the only evidence available to the ID that the demand in question is indeed being met, given that the EGO has no direct channel through which to communicate with it. As a result, such tensions, being indicative of action and demands-in-the-process-of-being-met, are experienced by the EGO as pleasurable, and desirable in themselves.

In this respect the ID acts like someone with remote control over something, such as a radio-controlled model. Here, as with the ID, there is an up-link to the model (the radio transmissions), but no down-link as such (the model does not transmit back). Instead, the controller has to observe the model's behaviour and respond accordingly, so that if he signals a left turn, the appearance of the model turning left will satisfy him that his command has been received and acted upon. However – and again perhaps just like the ID – failure to respond usually solicits further, and more urgent commands, repeated until the system responds.

In other words, my point is that unfulfilled fitness-maximizing demands, not drives as such, are experienced as unpleasurable, whereas demands being met are experienced as pleasurable. To use another analogy, one might say that the situation is rather like someone faced with demands from the tax collector. No one likes receiving tax demands – like Freud's ego in regard to the id, they represent costs registered as unpleasure. But being able to pay such a demand can be gratifying in the sense that one is relieved to be able to meet it, rather than not.

In short, a tax demand is one thing, paying it is another. What matters is not the paying as such, but whether I have enough funds to meet the demand. If I do, all is well and good; if not, I worry, not so much about paying in itself (after all, I am happy to pay for things I enjoy), but about my ability or otherwise to meet the demand made on me. Similarly with my concept of the EGO. Initiating tension-increasing activity is not unpleasurable to it as long as the basic demands of the ID are being met. Only if the latter cannot be met will the EGO experience unpleasure, tension or actual pain. This is because the EGO conceived here is regarded as an evolutionary adaptation ultimately charged with carrying out tactics to satisfy the grand, fitness-maximizing strategy of the ID. When it succeeds it is rewarded with pleasure, but is punished with pain and anguish when it fails to satisfy interests which are ultimately not so much those of the individual EGO as the genes which constitute the biological, evolutionary foundation of the ID.

An illustration of this can be seen in dreams. According to Freud, every dream is the fulfilment of a wish, and. overwhelming observational evidence bears this out.[8] But, if the line of enquiry suggested here is pursued, we might go a little further and say that,

---

[8] For a concise account of Freud's theory of dreams, with examples, see Badcock, *Essential Freud*, chapter 3.

not only is every dream the fulfilment of a wish, but that *every dream is the fulfilment of a fitness-maximizing wish.*[9]

It seems entirely in accordance with Freud's findings and with observation that no one dreams about wishes which have been thoroughly gratified in waking life. It seems that dreams are the EGO's way of disposing of inclusive-fitness-maximizing demands which it has not been able to meet adequately at the time they were perceived. If this view of the matter is correct, unrealized desires must leave an unpleasurable tension attached to that particular demand and to the EGO's failure to gratify it.

Although the EGO's motto may be, 'I am, therefore I act', the exigencies of real life mean that it is not always able to do so. It seems that, following a day on which it experienced some limitation on its freedom to act in accordance with fitness-maximizing demands, it might resort to a secondary tactic, one enshrined in the motto, 'If I cannot act, I can at least dream!' By dreaming, the EGO attempts to dispose of the unfulfilled demand by fantasy means open to it when action in the real world is inhibited by the state of sleep. This reduces tension, clears that particular ungratified demand and frees the EGO to meet the new demands of the next day.

Such may be the value of dreaming from the evolutionary and biological point of view. Certainly, such an interpretation would fit neatly with the classical topographical, dynamic and quantitative viewpoints enshrined in Freud's original theory and would serve to complement and complete them in the manner which I am advocating as generally desirable for psychoanalysis.

Freud's finding that so many dreams were motivated by sexual wishes may have scandalized his prudish contemporaries, but it should not have surprised anyone who had read their Darwin and realized what it meant. If evolution is directed by differential reproductive success – and this, as we have seen, is the essence of Darwin's theory – then Freud's finding with regard to the wishes revealed by dreams is entirely predictable and just as it should have been. It also leads us directly and naturally to the principal topic of this book: the evolution of human behaviour seen from the perspective of a theory which unifies the Darwinian and Freudian theories of sex.

[9] Recently an ingenious attempt has been made to find an evolutionary, adaptive dimension to dreams, at least as they can be reported to others (A. T. Lloyd, *On the Evolution of Instincts: Implications for Psychoanalysis*).

Three Essays on a
New Theory of Sex

1

Sex and Parental Investment

Cost and benefit in the analysis of sex

Between 1905, the publication date of Sigmund Freud's *Three Essays on the Theory of Sexuality*, and well into the 1960s psychoanalysis could claim a prominent place in – if not a virtual monopoly on – leading ideas about human sexual behaviour. This was despite the fact that its findings were often ignored by non-analysts, who preferred the more readily available findings contained in Kinsey-style surveys or journalistic accounts to the relatively inaccessible, if nevertheless vast, accumulation of clinical psychoanalytic observations on the subject. Also, dubiousness about the basic theory may have made many outside the narrow and rather exclusive world of psychoanalysis doubt the value of its findings.

Whatever the truth about psychoanalytic insights into what Freud, writing at the end of his life, could still call 'the great riddle of sex',[1] it is no longer possible to defend the view that psychoanalysis remains at the forefront of the subject. Since the mid-1960s that privilege has decisively and perhaps permanently been accorded to sociobiology: a term which describes the modern, Darwinian study of the evolution of social behaviour. Perhaps it is not entirely surprising that this should be the case; Freud himself had ended his Essays with the comment that 'the unsatisfactory conclusion ... that emerges from these investigations of the disturbances of sexual life is that we know far too little of the biological processes constituting the essence of sexuality to be able to construct from our fragmentary information a theory adequate to the understanding alike of normal and of pathological conditions.'[2]

[1] Freud, *Analysis Terminable and Interminable*, XXIII, 252.
[2] Freud, *Three Essays*, VII, 243.

17

At the time of writing the preface to the third edition of the book he could say that 'the present work is ... deliberately independent of biology' and that

> I have carefully avoided introducing any preconceptions, whether derived from general sexual biology or from that of particular animal species ... my aim has rather been to discover how far psychological investigation can throw light upon the biology of the sexual life of man. It was legitimate for me to indicate points of contact and agreement which came to light during my investigation, but there was no need for me to be diverted from my course if the psychoanalytic method led in a number of important respects to opinions and findings which differed largely from those based on biological con-siderations.

However true that may have been in 1914, it seems that today the classical Freudian view of sexuality differs much less from the modern biological one than might have been supposed. This is mainly thanks to the findings of sociobiology, which has cast a surprising new light on the whole question of sexual behaviour and, as I hope to be able to show, has created the possibility of a synthesis of biological and psychoanalytic insights into human sexuality which was impossible to foresee when Freud wrote his *Three Essays*.

In large part, and as I argued at length in the book which forms a parallel, complementary study to this one, sociobiology began with the demise of what has become known as *group-selection* – the belief that natural selection operates at the level of the group or even the entire species.[3] This contention seemed to make self-sacrifice and altruism in general self-explanatory, because they were evidently in the interests of the group, or, at the very least, of others apart from the altruist. In sexual behaviour group-selection also seemed to be an obvious concept – after all, sex was all about reproduction, and reproduction was for the benefit of the species!

Nevertheless, and true as this might seem at first sight, certain problems remained. Any man foolish enough to use such a rational-ization in the course of seduction – something along the lines of 'let's do this for the benefit of reproducing the species' – would almost certainly find his reasoning less than compelling if the lady

[3] Badcock, *The Problem of Altruism*, introduction.

in question were considering the potential personal cost to herself of the action he was contemplating. Perhaps philanderers do sometimes excuse themselves with the observation that they are selflessly performing a service for the future of the human race in fathering children on many different women. But no one else is likely to take such a complacent view – especially those whom they then abandon to do their rather more exacting and extensive service to human reproduction by way of actually bearing their offspring and bringing them up.

Nowhere more than in the context of sexual behaviour is the shallowness and falsity of group-selectionism revealed; but this is only so if one considers the personal interests of the individuals directly concerned, something that sociologists and social anthropologists have notoriously not done. Taking instead the view that only the group, and not the individual, is what matters in human affairs, they have tended to see sex and sociability as almost synonymous as a result of the naive and misleading observation that, because 'it takes two to tango', the coming together of the sexes for purposes of reproduction is the natural foundation of the family in particular, and of the wider society in general.

This was never the view of Freud who, perhaps because of his more personal concerns and purely individually-based method of research, tended to be sensitive to the potential conflicts involved, both between individuals and within them. With the realization that natural selection operated not at the level of the family, group or species, but at that of the individual and even the individual gene, sociobiology came to take a similar view. One of its very first advocates was to state the principle explicitly when, in words which might have come from the writings of Freud, he affirmed that 'sex is an anti-social force in evolution.'[4]

To understand why this is so and why sociobiology took this individualistic, dynamic view of sex and its relation to social behaviour we have to consider one of its most important ideas: the theory of parental investment.

This might be best approached by considering what is perhaps the most fundamental question about sex, namely, what constitutes the difference between the sexes? The modern, biological answer to this question is simple and unambiguous: *the size of the sex cells in*

[4] E. O. Wilson, *Sociobiology*, p. 314.

question. Almost universally, one sex produces a relatively large, relatively immobile sex cell (or *gamete*), always provided with stored nutrients; the other produces a relatively minute, relatively mobile sex cell without any such provision. Where this is so, the sex with the larger cell is the *female*, that with the smaller the *male* (in those rare cases where such *gametic dimorphism* is not found and sex cells are of identical appearance, one is arbitrarily labelled *plus* and the other, *minus*).

According to the most widely accepted modern theory, disparity in the size of the gametes probably developed from a situation of near equality between them (rather like the 'plus' and 'minus' situation mentioned above). Modelling of the interaction of such nearly equally sized sex cells suggests that genetic variation and natural selection would tend to drive them to the extremes, so that ultimately only two stable configurations remained which could always exactly complement one another: a tiny, mobile male sex cell and a larger, less mobile, female one.[5]

In the case of human beings, for instance, the male sperm is the body's smallest cell and the female ovum the largest; but even this great discrepancy in size underestimates the true difference, since it fails to take into account the extensive provision which the human female has to make after fertilization. A better contrast might be that between the sperm of a cock and a hen's egg, because in this case the egg does indeed contain all the nutrients required to bring the chick to a live birth, and such a contrast spans many orders of magnitude – from the microscopic to the distinctly and tangibly macroscopic. To make the disparity in size even more graphic and easy to visualize, imagine that the roughly egg-shaped head of a cock's sperm were expanded to equal the size of an average hen's egg. The hen's egg on the same scale would have a long axis of more than one kilometre, or just under three-quarters of a mile!

The essential difference between male and female, then, can be described in terms of *parental investment*, initially understood as the amount invested in the production of the sex cells. However, sexual reproduction often involves further provisioning and care of the developing young (for instance, in mammals via the placenta and breast), which means that the concept of parental investment cannot

[5] G. A. Parker, R. Baker & V. Smith, 'The Origin and Evolution of Gamete Dimorphism and the Male–female Phenomenon'.

be limited to the production of sex cells alone. It must be extended to include all provisioning, transportation, protection, care, education and rearing of the young by whichever parent carries it out and is defined as *any behaviour which increases the reproductive success of any of an organism's offspring at the expense of the remainder of the parent's reproductive success.*[6]

We can now begin to see that one virtue of the theory of parental investment as applied to sex is that it concentrates attention on the costs, as well as the benefits involved. The old-fashioned group-selectionist approach concentrated attention on the benefit of sex for the family, group or species, without much attention to its liabilities and with no indication of the differential costs to the individuals concerned. Sex, reproduction and the rearing of off-spring were good for the family, group or species and, in terms of ultimate benefit to the continuation of life, self-evidently valuable. The theory of parental investment, however, also takes into account the costs to individuals concerned; and here it has found a notable discrepancy: in general, and with astonishingly few real exceptions, *females invest much more in offspring than do males.*

Most sociologists and social anthropologists have been male and, in the case of the former, largely reliant on survey or statistical data which stress the normal, average or cumulative trends over the individual viewpoint, subordinate tendency or deviant fact. Most social anthropologists have relied largely for information, translation and support on other males, usually prominent and established men in the societies they studied, and seldom had access to the views of women, children or other, subordinate individuals. As Warren Shapiro has pointed out, the 'Ego' of kinship studies is always 'the male Ego'[7].

When we add to this the fact that anthropologists are not usually able to speak the native languages of the societies they study with sufficient fluency and accuracy to pick up undertones and nuances of discontent, dissension or disaffection we begin to see why the accounts which such they subsequently publish reflect a notably complacent, benefit-orientated view of sex and the family. Not surprisingly, it is one which stresses the normal as opposed to the

[6] R. Trivers, 'Parental Investment and Sexual Selection' Cf. R. A. Fisher, *The Genetical Theory of Natural Selection*, pp. 142–3.
[7] W. Shapiro, *Miwuyt Marriage*, p. 13.

deviant, the holistic view as opposed to the interests of the individual, the official system as opposed to the implicit reality.

Considerations such as these suggest good reasons why traditional social science has tended to identify with the 'obvious', 'official', 'collective' view of sex as synonymous with sociability. According to this view sex is at worst something which 'cannot safely be left without restraints' because it is 'a powerful impulse, often pressing individuals to behaviour disruptive of the cooperative relationships upon which human social life rests'.[8] At best, 'sex is a supremely social drive in which the libido is harnessed in service of society.'[9]

An excellent example of the kind of distortions and, indeed, absurdities to which this view of sexuality can lead is provided by the works of the influential anthropologist, Claude Lévi-Strauss. Purporting to demonstrate that kinship systems are based on the reciprocal – and therefore apparently equitable – exchange of women between groups of men, he gives an account of polygamy which tries to excuse the obvious inequality which it produces in the distribution of wives by reference to the value of polygamous chiefs 'not to the particular individuals whose sisters or daughters he has married, and not even to those perhaps condemned for ever to celibacy by the exercise of his polygamous right, *but to the group as a group,* for it is the latter which has suspended the common law in his favour.'[10] How well the chiefs in question would be pleased by his next assertion to the effect that 'polygamy, therefore, does not run counter to the demand for an equitable distribution of women'! To which he perhaps ought to have added that in the number of their wives, therefore, all men are equal, but some are more equal than others!

Nevertheless, one might well still wonder what those not fortunate enough to have wives because of this 'equitable' distribution of women actually think about it. Lévi-Strauss gives no evidence of knowing, or indeed caring. Their personal interests have been smothered in the alleged collective interests of the group – so conveniently identical to those of the ruling minority. In a manner typical of the modern social sciences, some alleged benefit to the group is used to discount the real costs incurred by some to pay for the privileges enjoyed by others.

[8] G. P. Murdock, *Social Structure*, p. 4.
[9] T. Gregor, *Anxious Pleasures: The Sexual Lives of an Amazonian People*, p. 68.
[10] C. Lévi-Strauss, *The Elementary Structures of Kinship*, p. 44; my italics.

Unlike the social sciences, classical psychoanalysis has from the beginning listened, with all the lack of prejudice and preconception which its characteristic method of free association allowed, to the private, personal views of women (who at the beginning constituted the majority of its patients), men (not all of whom were socially dominant, conventional or respectable individuals) and, at least since the 1920s, to any child old enough to speak coherently. As sociobiology was later to find, classical psychoanalysis soon realized that sex was not synonymous with sociability, harmony and consensus, but that, on the contrary, it was deeply involved with conflict, aggression and all kinds of disharmony, not only within the group, but also within the individual.

In saying all this I have had to qualify my use of the term 'psychoanalysis' with the adjectives 'classical' or 'Freudian' in order to indicate that I do not mean to include the post-Freudian, neo-Freudian or pseudo-Freudian versions of psychoanalysis which have proliferated bewilderingly since the 1920s. In the past I have been roundly condemned by my critics for my insistence on Sigmund and Anna Freud's version of psychoanalysis as the only one worth discussing. In the light of the foregoing the reason for this becomes clear: it is simply that most other interpretations of psychoanalysis have defaulted on the very aspect of it which I am emphasizing as especially valuable. In other words, whereas Freudian psychoanalysis remained resolutely individualistic and put the libido theory centre-stage, most other schools of psychoanalysis and variants of it have tended to surrender to contemporary prejudices and to emphasize the social rather than the individual, symbolic 'gender' rather than biological sexuality.

In this respect they have been entirely in harmony with modern sociology and this might explain why the so-called 'object relations' school has become so influential. To the extent that this pervasive climate of thought has tended to emphasize early social relationships rather than primal libidinal attachments, it might be more accurately termed the '*social relations*' school. Furthermore, the almost total disregard of biological determinants of behaviour found in nearly all modern versions of psychoanalysis shows that, even if true of Freud himself and still true of those who adhere to his classical formulations, the comments above cannot possibly include most of what passes for 'psychoanalysis' today.

An example of this is provided by Robert Stoller's *Presentations of*

Gender. Not only does the approach of this book appear cultural-determinist to the point where its author claims that 'in most instances in humans, postnatal experiences can modify and sometimes overpower already present biologic tendencies,' he concludes that the 'Sambia' of New Guinea (this is not their real name) 'need ... heterosexuality for producing future communities'.[11] Presumably this is for the same kind of reason that Lévi-Strauss thinks that certain tribal groups 'need' their chiefs to have sexual privileges denied to ordinary mortals: in other words, *for the good of the group.* When Stoller speaks of the 'cultural mechanisms through which Sambian masculinity was created' we are left in no doubt that here so-called 'psychoanalysis' and orthodox cultural-determinist sociology have become almost one and the same thing. 'Sex' has become the subjective concept of 'gender'[12] and does not exist for the reproduction of the individual's genes, but is reproduced itself by the culture, for the culture and through the culture. I mention Stoller in particular because, as we shall see later, his case-material, although interpreted by him in a predominantly cultural-determinist manner, takes on a quite new significance when seen in terms of the theory of parental investment.

In this respect what has befallen Freud seems closely comparable to what befell Darwin. In the latter's case too it was true that, for the best part of a hundred years after the appearance of his theory, much confusion existed about what it really meant and what passed itself off as 'Darwinism' was often only very dubiously Darwinian at all. Indeed, these comments apply to Darwin himself and explain the otherwise startling paradox that the editions of his books now read are not usually the last to appear in his lifetime, but the first.

This is a state of affairs which runs clean counter to the naive official view of scientific progress as an ever-on-and-upward movement without regressions, deviations or periods of latency. Rather it suggests the contrary and much more realistic view: namely, that Darwin himself was hindered in the development of his insights by contemporary criticism, ignorance and prejudice to the point that his own later versions of his ideas now seem less valuable than his first, less inhibited and compromised formulations. Admittedly, Mendelian genetics was integrated with Darwinism in the 1920s and

---

[11] R. J. Stoller, *Presentations of Gender*, pp. 6 and 190 (the latter co-authored with G. H. Herdt.)
[12] Ibid., p. 11.

1930s and the foundations of later important developments were laid. But, by and large and with remarkably few real exceptions, Darwinism, like Freudianism today, seemed to becom stymied for the greater part of its first century.[13]

Excepting the problem with genetics, there appear to have been two principal reasons for this. One of these was confusion about the meaning of the term 'fitness' and the tendency to see it in terms of bodily health and welfare, rather than *reproductive success*, which is how it is seen today. The other was the tendency already mentioned to discount the individualism and reductionism of Darwin's original formulation of his theory in favour of ideas regarding *group* rather than *individual* selection. As I remarked earlier, this seemed to make social behaviour self-explanatory and to make sex look like something which was 'for the good of the species', rather than in the interests of the individual.[14]

In other words, what inhibited the development of Darwinism with regard to understanding the evolution of sexual and social behaviour was exactly what has halted genuine progress in psycho-analysis: an emphasis on group, rather than individual, interests and a fudging of the always controversial issue of sexuality and its determining role. Once modern Darwinism purged itself of both group-selection and confusions about 'fitness' and came to see evolution as driven by the differential reproductive success of indi-viduals and their genes, enormous advances became possible, ad-vances which would have been possible long before, had those obstacles to progress not been present.

My aim in this book is to begin to bring about something similar with regard to Freud and psychoanalysis, and if I seem to ignore much of what passes for psychoanalysis and Freudianism today, it is for exactly the same reason that modern evolutionary theory completely ignores the Social Darwinism of earlier times and its badly contaminated categories, confused thinking and incorrect assumptions. If I am criticized for sticking too closely to the originals in psychoanalysis my reasons are much the same as those which have resulted in the modern editions of Darwin's works being reprints of the first, not of the later and much revised editions

[13] For a crucial example see Robert Trivers's discussion of how Haldane just missed the insight which was to revolutionize modern Darwinism in his *Social Evolution*, p. 46.

[14] Badcock, *The Problem of Altruism*, introduction.

which appeared towards the end of his life.

My justification for this is that scientific progress is not as simple and as straightforward as is commonly supposed; on the contrary, real regressions, periods of confusion and temporary reversals are often possible, especially in the immediate aftermath of a traumatic scientific revolution. Following one of these periods of upheaval, getting back to the original inspiration at its root is not necessarily evidence of conservatism or reaction, but can sometimes be a true sign of progress.

In my view 'object relations' theory and most of the other modern fashions in recent psychoanalysis will go the way of the Social Darwinism and pseudo-Darwinisms of the past. They will be superseded in psychoanalysis for the same reasons that they were superseded in evolutionary biology: because they confused the issue about the level at which explanation should be sought and because they prudishly ignored the fateful, determining and biologically based imperative of sexuality and the final arbiter of evolutionary change which it drives: differential reproductive success.

Sexual strategies

The general conclusion must be that sex and social cooperation are by no means as closely and unproblematically related as might have been supposed, either within the family or in the wider society. On the contrary, once both the costs and the benefits of sexual behaviour to the individuals concerned are considered in the light of our modern understanding of evolution as operating at the level of the individual and the gene we can begin to see that sex is indeed a predominantly anti-social force, both within evolution as a whole and within the individuals on which it acts.

Even in cases where there is an apparently equitable division of labour between the sexes in the rearing of offspring, occasions for conflict and ambivalence can occur. For instance, most birds are monogamous, with some even manifesting life-time adherence to one partner, and this makes them seem something of a paradigm for monogamy in biology. The fact that the majority of birds is monogamous is generally thought to reflect the fact that they need to make considerable provision for their offspring in the form of food and nests and that such extensive investment normally requires the participation of both parents.

But in those species of monogamous birds whose mating period is sufficiently long, circumstances can arise where, once chicks are mature enough to be fed by only one parent, the other might desert its first mate in time to find a second and thereby double its reproductive success in that particular season. When we recall that evolution is directed by such differential reproductive success we can begin to see that desertion might be rewarded because individuals with genes for successful desertion might leave twice as many offspring as those without. In these conditions desertion might never be a dominant reproductive tactic, but it might nevertheless be an important subordinate one and would certainly be one which would illustrate the naivete of believing that just because the species was monogamous males and females would always rear their offspring in perfect harmony.[15]

In the example above either sex may in principle desert the other; what matters is only the timing: too early desertion may threaten the survival of the first brood, too late may allow the other partner to desert first. But in practice this is not always the case. More usually it is males who can afford to be the sexual adventurers, while females pursue a strategy which is more coy. In Darwin's words, 'males are almost always the wooers,'[16] and in Freud's, "Masculine" and "feminine" are used sometimes in the sense of activity and passivity,' and it is this distinction between the sexes which 'is the essential one and most serviceable to psychoanalysis'.[17]

It is also a distinction easily observed in human behaviour, and one which is graphically illustrated by this account of a society far removed from the 'Victorian' cultural values which are alleged to have influenced both Freud and Darwin:

Mehinaku society comes as close to a stereotypical male sexual utopia as is possible in a functioning family-based community. Married men are able to have noninvolving sexual relationships with many women with little risk to their reputation or community standing. Even in this setting, however, male libido expands beyond the boundaries of available sex. The men complain that 'women are stingy with their

[15] Trivers, *Social Evolution*, pp. 267–9.
[16] C. Darwin, *The Descent of Man and Selection in Relation to Sex*, p. 397.
[17] Freud, *Three Essays*, VII, 219n.

genitals.' ... The imbalance of gratification and desire is perceived by
the villagers as a powerful force. Men are assumed to be perpetually
sexual.[18]

The reason for this state of affairs, which is evidently widespread
and probably universal, returns us directly to our initial consideration of the essential difference between the sexes. Because the sex
cells of the male are, by definition, always much smaller, much
more numerous and much more mobile than those of the female,
the male will usually be found to be rather different in behaviour to
the female. Whereas she, with her greater degree of investment in
her sex cells (possibly enhanced by internal gestation or brood-care
along with provisioning of the young) needs to be discriminating
about her potential sexual partners – 'stingy with her genitals', in
the graphic words of the Mehinaku – he need not. On the contrary,
if the male's investment in the young terminates with fertilization
his effort is likely to be mainly directed towards fertilizing as many
females as possible. This is because, here again, natural selection
favours reproductive success and a male who leaves more offspring
will pass the genetic determinants for that behaviour on to more
descendants than one who leaves fewer.

Such a male might still, of course, have to make considerable
reproductive effort; but his exertions in this direction are more
likely to be directed towards courtship and gaining access to females
than to providing parental care.[19] Here the effort involved can be
very considerable indeed, especially if gaining access to females
involves competition with other males, similarly bent on careers of
indiscriminate mass fertilization. In the case of the Mehinaku,

Men who do not get along explain their hostility in terms of sexual
jealousy. Men may even seek revenge against thieves and gossips by
having sex with the culprit's wife. Allegations of witchcraft often
derive from the tension of competing for the same women. Few
Mehinaku relationships or institutions escape the tensions generated
by sexual desire and frustration.[20]

[18] Gregor, *Anxious Pleasures*, pp. 200–1.
[19] Nevertheless, as we shall see shortly, the two are by no means exclusive. It is
just that for present purposes we are trying to distinguish the sexes in the most
obvious and simple manner.
[20] Gregor, *Anxious Pleasures*, p. 201.

One notable external sign of such antagonistic male behaviour is *sexual dimorphism*. This simply means a consistent difference in bodily appearance between the sexes, consisting both of sex-specific variations in common attributes such as weight and size, and secondary sexual characteristics like beards or antlers. Where males come into direct physical conflict over access to females attributes like body build, canine teeth, horns or other weapons may be selected because they confer advantages in such conflicts on their possessors. Where *female choice* is more important, other adaptations like elaborate or highly coloured plumage may play a sex-defining role.

Although Darwin originally distinguished between natural and sexual selection, invoking the latter to explain the adaptations just mentioned, modern evolutionary theory tends to treat the two as much the same because in either case it is eventual reproductive success which counts. Natural and sexual selection only seem to be different – or even to come into actual conflict – if one adopts the fallacious view that selection operates on groups, rather than individuals. Since we have already seen that this is not so but that, on the contrary, selection operates at the level of the individual organism and even at that of the individual gene, the individual's interest in its own reproductive success must mean that natural and sexual selection are ultimately one and the same thing.

Again, the erroneous view that 'fitness' (as in 'survival of the fittest') means personal health rather than reproductive success encourages the belief that where sexual competition or female choice exact costs in terms of personal fitness, the two principles of selection are in conflict. Peacocks probably do pay a price in terms of personal survival for their wonderful tails – for instance, in their reduced ability to escape predators; but this must be compensated for by the enhanced sexual success which such ornaments procure. In the last analysis both sexual and natural selection mean the ultimate reproductive success of the genes for which the individual organism is no more than a temporary 'packaging' or container. Looked at from this point of view, natural and sexual selection merely reflect a difference in emphasis with regard to the adaptation concerned, rather than a conflict relating to the principles of selection involved.

The female will tend to evolve rather differently from the male because, although in exceptional cases competition among females for desirable males may be the rule, females will normally be more concerned with their investment in offspring and with that of their mate, if it should occur. This means that females should evolve to be more coy than males in the sense that they should be more conditioned to the costs of potential matings and might even exercise a discriminating choice if given the opportunity. Where the male parent does contribute to the care of the offspring such discrimination would probably relate to his desirability and trustworthiness as a provider as much as to anything else. But even if male parental investment is not at issue female choice can still be an important factor. In numerous animal species where males are decorated, rather than merely armed, Darwin concluded that a male's attractiveness to females seemed to be the decisive factor and that it indicated that females were a resource not merely to be fought over by males, but to be won over by them.

In the past, and doubtless under the influence of views which ignored individual choice and emphasized group benefit, female choice tended to be overlooked. Females were regarded as obliged to submit to male advances 'for the good of the species'. Since the males who mated with them were often socially dominant, a situation existed exactly analogous to that in the human social sciences. Females were presumed to go along with dominant male interests without much protest. However, modern field studies, even of animals where males do indeed seem to be sexually dominant, have begun to reveal the extent to which females do in fact exercise choice.

For instance, in the case of many quadrupedal mammals the male mounts the female, who appears to submit passively to his advances. Yet animal breeders know that one of the main problems with controlled mating is the difficulty in making the female stand still. Where a quadrupedal male has to mount a female in order to fertilize her, it only needs the female to move or walk away for all his efforts to come to nothing! This is why even highly aggressive males with vastly superior armament in the form of horns or antlers like sheep or deer nevertheless court their females gently, knowing that the cooperation of the female in question is essential if mating is to be successful. Furthermore, it seems a safe prediction that as more careful and individualistic field studies are made across the entire spectrum of animal species female choice will be revealed to be a more important factor than might once have been supposed when female interests were uncritically assumed to be identical with those of males.

In a manner which now seems far ahead of his times, Darwin had no hesitation in applying the principle of female choice to human beings. He seems to have taken a rather less prejudiced view than many modern social scientists when he observed that

> with savages the women are not in quite so abject a state in relation to marriage as has often been supposed. They can tempt the men whom they prefer, and can sometimes reject those whom they dislike, either before or after marriage.[21]

By contrast to influential modern social theorists like Claude Lévi-Strauss, Darwin did not see women as mere arbitrary 'signs' to be exchanged in a system of kinship 'messages' among groups of men. Like classical psychoanalysis, but unlike much modern social science, evolution from the beginning took female choice seriously and considered more than merely the interests of powerful males in the determination of human affairs. Darwin's belief in the reality of female choice was ridiculed at the time he put it forward and was assumed by later writers to be a weak feature existing for the benefit of the species (for instance, to avoid 'wasted' matings with males of the wrong species). If emphasis on the role of the individual made Darwin's and Freud's view objectionable to many, their respective insights into female interests made them even more absurd in the eyes of the dominant male orthodoxy of their times. Indeed, the controversial nature of the concept of female choice even today may, for all we know, reflect a similar prejudice on the part of some.[22]

One consequence of the concept of female choice which was immediately apparent to Darwin was that where it was an important factor, females had a leverage on the direction of male evolution denied almost entirely to males in relation to females. As he put it in the pages of *The Descent of Man*, 'preference on the part of women, steadily acting in any one direction, would ultimately affect the character of the tribe; for the women would generally choose not merely the handsomer men, according to standard of

---

[21] Darwin, *The Descent of Man*, p. 374.
[22] For an excellent modern discussion of female choice, see Trivers, *Social Evolution*, pp. 331–60.

taste, but those who were at the same time best able to defend and support them.'[23]

This conclusion is a logical consequence of our original consideration of the nature of the difference between the sexes. If males are generally much less discriminating than females, their effect on female attractiveness may be very slight: since they are so undemanding in what they regard as desirable, more or less anything will do. But for the 'choosy' female, and especially for the 'really choosy' female, by no means anything will do. On the contrary, she will direct the evolution of male characteristics if her choosiness about them promotes the reproductive success of her offspring, which will include, of course, those females among her offspring who inherit her discriminating trait. The consequence, as Darwin saw, is the tail of the peacock, the building activity of bower birds, and the whole astonishing phenomenon of male display wherever it is found.

Furthermore, as I shall try to show later, female choice may even be the decisive factor in recent human evolution and may account, not merely for some of the most notable features of the physical form of both sexes, but also for many of our unique behavioural and psychological adaptations. Indeed, it may well be at the root of the greatest mystery of human evolution: the vast expansion of the human brain and the correspondingly enormous enhancement of the human behavioural repertoire which we take to be the unique evolutionary achievement of our species.[24]

Males, in contrast to females, are expected to be much less discriminating, sometimes to the point of mating with partners which are not even members of their own species. In his *Three Essays on the Theory of Sexuality*, Freud had pointed out that 'sexual intercourse with animals ... is by no means rare, especially among country people'[25] and he could also have added that, although sexual relations between women and animals are by no means unknown, men are by far the more likely to be involved.

This fact stands in sharp contrast to Edward Westermarck's view that 'a real, powerful instinct' guards human beings from intercourse with animals or close relatives and that such matings are

[23] Darwin, *The Descent of Man*, pp. 374–5.
[24] See pp. 142–186 below.
[25] Freud, *Three Essays*, VII, 148.

selected against by evolution 'for the good of the species'.[26] Here once again, Freudian psychoanalysis and modern biology seem to be in agreement in noticing that, if the costs to individuals of sexual activity are taken into account, the presumed benefit to the species of avoiding apparently undesirable matings becomes much less 'natural' than it might at first appear.

Indeed, this is an observation which is widely true and is by no means confined to human, or even to animal cases. Certain species of orchid, for instance, exploit the low cost of matings to males and their consequent lack of discrimination by displaying flowers which resemble the female of a particular kind of wasp and emitting a perfume which is similar to the female's sex pheromone. Easily duped, males attempt to copulate with the flower and, in so doing, carry off its pollen. 'Among invertebrates as diverse as butterflies and hermit crabs males are apt to court an astonishing variety of objects, indeed almost anything that bears some resemblance to a female.'[27]

Indeed, the case of the orchid which mimics the wasp should remind us that those plants and animals which rely on intermediaries to carry out fertilization for them almost always transfer the male, rather than the female sex cell. Such strategies are, of course, only refinements of an even more indiscriminate method of mating: the general broadcast of pollen to the winds or sperm to the waves seen in plants such as grasses or animals such as coral polyps.

Looked at from this point of view, the use of animate or inanimate intermediaries as an essential part of the process of reproduction represents a kind of fetishism. Although this may seem a somewhat strained use of the term, the fact that chimpanzees and baboons (albeit captive ones) have been reported to have adopted rubber boots for purposes of masturbation suggests that, even if more strictly defined, fetishism is not necessarily unique to man.[28] On the contrary, the fact that the vast majority of human fetishists are men rather than women, suggests that their biological masculinity may have something to do with their apparently 'perverse' behaviour and that sexual perversions in general

[26] E. Westermarck, *The History of Human Marriage*, p. 353.
[27] M. Daly and M. Wilson, *Sex, Evolution and Behavior* (2nd edn), p. 114.
[28] A. W. Epstein, 'Fetishism: A Comprehensive View'; 'The Fetish Object: Phylogenetic Considerations'; 'The Phylogenetics of Fetishism'.

may not be as inexplicable from a modern, individualistic evolutionary perspective as they may have seemed from the old-
fashioned, group-selectionist one.

Again, masturbation might have seemed unnatural from the
older point of view because it did not lead directly to reproduction
and was therefore a personal 'vice', anti-social and contrary to the
interests of the family, group or species. However that may be, wild
mountain rams 'may ejaculate spontaneously throughout the year
during interactions with other rams, while courting estrous or non-
estrous ewes, or after rising in the morning'. In goats 'the penis is
taken into the muzzle', as it is in ibex where 'ejaculation is a
smooth, quick operation in which the male puts the penis quickly
into his mouth, maybe using the cheek cavity as an artificial
vagina'.[29] Although this is an unusual (not to mention difficult)
method of masturbation in human males, it is by no means
unknown and, according to Kinsey's findings, most men try it out at
some time or another.[30] Male dolphins seem to be almost as inventive
in finding techniques for self-stimulation as do human males and
female elephants in single-sex groups have been seen to use their
trunks for purposes of mutual masturbation.[31]

Although we shall have to leave the whole question of possible
evolutionary foundations of human sexual aberrations until later,
suffice it to say in passing that the mere fact of being male must
often predispose an organism to be much less selective in its sexual
objects than would have been the case had it been female, and that
this lack of selectivity might on occasion extend to having no sexual
object at all save the self. All in all, and with only a partial exception
in the case of masturbation, it seems that we are justified in
concluding that 'women are not represented in most of the diagnoses
of perversion; whatever their fantasies, women practice perversions
less than men do.'[32]

Nevertheless, a male's lack of discrimination can carry indirect
costs, for not only does it result in wasted matings but, in the case
of the wonderful guppy, for instance, it also results in female
discrimination between males of similar species, producing male
guppies with characteristically decorated tails which advertise, so to

[29] V. Geist, *Mountain Sheep: A Study in Behavior and Evolution*, pp. 141 and 224.
[30] W. B. Pomeroy, 'Kinsey and the Institute', p. 44.
[31] R. H. Denniston, 'Ambisexuality in Animals', pp. 37–8.
[32] Stoller, *Presentations of Gender*, p. 9.

speak, their exact species. Since these tails may also make their possessors more vulnerable to predators, it seems that evolution has exacted an indirect, but real cost on the males in question and one which, paradoxically, results directly from the relative 'cheapness' of male parental investment as compared to female in the species in question (which, like mammals, undertake internal gestation of the young).[33]

The interplay of male and female interests in furthering their respective eventual reproductive success in the environmental conditions in which they find themselves will produce a particular, or range of particular *mating strategies*. In general, these can be characterized as *monogamous* or *polygamous*: the former being defined as one male mating with one female, the latter as one male with a number of females, which we may more specifically call *polygyny*, or the converse, one female with a number of males, which is known as *polyandry*. Finally, a number of females can be mated to a number of males, and this we may call *promiscuity*.

Sexual dimorphism, along with the development of specific secondary sexual characteristics pertaining to combat with other males, is typically associated with polygyny and promiscuity for the obvious reason that in both of these mating systems males may come into conflict for access to females. Polyandry is relatively rare, probably because only one male is normally necessary to fertilize a female and where male parental investment is concerned, natural selection will not reward males who invest in other males' offspring. Where polyandry does exceptionally occur among human beings, the woman's husbands are usually brothers, so that whatever sacrifice they may personally make is partly compensated for by considerations of inclusive fitness – personal sacrifice for benefit of near relatives who share, in this case, half one's genes.

In the context of monogamy, by contrast, sexual dimorphism is likely to be less pronounced and male secondary sexual characteristics pertaining to inter-male conflict not usually so elaborately evolved because of the normally high cost of these adaptations to their possessors. The sexes will tend to resemble one another in body-build, size and external appearance and will often be found to cooperate in parental care of the young. For instance, the majority of birds are monogamous and sometimes non-dimorphic to the

[33] N. R. Liley, 'Ethological Isolating Mechanisms in Four Sympatric Species of Poeciliid Fishes'.

point of apparent identity of appearance between the sexes. This is believed to be related to the fact that fledglings usually need the efforts of both parents to provide them with the food which they need in order to grow to the point where they can fend for themselves.

Yet even in an ostensibly monogamous species it would be wrong to assume that fundamental sexual differences carried no weight or that even slight sexual dimorphism was insignificant. An ingenious experiment carried out on a species of monogamous swallow where the sexes are externally alike except for the two longest tail feathers being on average half an inch longer in the male showed that artificially lengthening or shortening the tail feathers of particular males dramatically affected their reproductive success. Males with shortened tails took four times longer to find mates than those with lengthened ones, who also raised two clutches more often and were more successful in polygamous matings with otherwise monogamously mated females.[34]

In this case it seems that male gallivanting and female choice of even slight dimorphic characteristics occur, suggesting that even in an ostensibly monogamous species evidence of polygamy can be found and that sexual strategies should not be regarded as fixed for any species in a simple or necessarily straightforward way. On the contrary, and as we shall see later, although one mating strategy may be dominant in a species, subordinate or deviant strategies can and often do exist (not least in human beings where, as we shall also see later, they take on the fascinating form of psychological deviations).

Cryptic sexuality

It is important to notice that in principle there is nothing to prevent sexual strategies becoming mixed, often with complex and paradoxical effects. Perhaps the best test of the theory of parental investment and its success in explaining sexual behaviour is the most paradoxical case of all: that of species in which the usual sex roles are reversed and males make a high degree of parental investment. In the case of sticklebacks, for instance, although the

[34] A. P. Møller, 'Female Choice Selects for Male Sexual Tail Ornaments in the Monogamous Swallow'.

female lays the eggs, the nest is made and guarded by the male. Competition among females for access to desirable males is in fact greater than antagonistic interactions between males: females in one study averaged thirty-eight aggressive encounters per hour, males only twenty-six.[35] Again, in the case of seahorses, males incubate the fertilized eggs in a pouch, leading to high courtship colouration and competition among females for access to desirable males. Heavy parental investment by males in the case of the Mormon cricket leads to 'coy', 'choosy' behaviour on their parts; and males of a species of moorhen who contribute 72 per cent of the incubation are fought over by females who exhibit considerable sexual dimorphism.[36]

Such apparent reversals of sexual role are by no means as uncommon as one might think and often seem 'perverse' and 'unnatural' by any standard – suggesting that they are also 'paradoxical' and inexplicable by any theory. An example taken from human behaviour is one of the earliest and arguably most important findings of psychoanalysis – what Freud called *bisexuality*. By this he understood the existence of psychological factors pertaining to one sex in a person of the other. Furthermore, although found to be a prime factor in the development and causation of neurotic conflict and – not surprisingly – especially notable in homosexuals, it was also perfectly normal in the sense that some evidence of it could be found in everybody, at least as a *latent* factor (and therefore one normally only accessible to psychoanalytic investigations). In effect this meant that the psychoanalytic method revealed sexual conflict not merely within the species, or the family, or during childhood, but within individuals themselves.

Although such a finding might at first seem as alien to the theory of parental investment as it was to earlier, cruder biological views of sex, here once again modern sociobiology has strongly endorsed the general finding regarding 'the widespread capacity in animals for the expression of behaviour appropriate to the opposite sex'. This capacity exists because 'the neural organization underlying behavior typical of one sex often lies dormant in the brain of the other sex too, and can be made to manifest itself by the right stimulus.'[37] For instance, in the case of a species of unisexual

---

[35] S. Li and D. Owings, 'Sexual Selection in the Three-spined Stickleback.'
[36] Trivers, *Social Evolution*, pp. 215–19.
[37] Daly and Wilson, *Sex, Evolution and Behavior*, p. 173.

whiptail lizard the fact that all animals are female does not prevent them engaging in male copulatory behaviour, in response not to a male hormone like testosterone, as one might expect, but to progesterone, a hormone normally more involved with the female sexual cycle.[38]

Again, on rare occasions females in certain species of fish can literally turn into males and transform their ovaries into testes producing genuine sperm. Sometimes this is a consequence of the female in question being the dominant member of a female group which has no male in it – something which might suggest that such a transformation of sex role could be seen as 'benefiting the group' or species. One can readily imagine a believer in group-selection seizing on this as an instance of the female transsexual converting to the other sex almost altruistically, as if it were to provide the species with males when the latter were in short supply. If this were the case, then it would be a notable exception to the rule that group-selection theories routinely ignore deviations and complications in sexual behaviour because they naively assume that, since sex is for the benefit of reproducing the species, only 'real', 'regular' males and females are required.

Nevertheless, if one looks at examples of sex-transformation like this from the point of view of the female in question one can readily explain it as in her (or his) personal interest, making such imputed benefits to the group or species quite unnecessary conclusions. For instance, where a dominant female transforms into a male it may well be because, given the greater variance of male reproductive success thanks to small, numerous and mobile sex cells, it will promote the transsexual's reproductive success rather considerably if he (or she) were to be a male, rather than a female.[39]

In this case being physically large and behaviourally dominant may equip a fish to be male better than to be female, but the exact opposite could occur and is reported. Cases are known where larger size is more compatible with being female and with producing eggs, rather than sperm, and here individual reproductive self-interest can produce the reverse sex change: male to female, rather than female to male as in the example above.[40]

[38] D. Crews, 'Courtship in Unisexual Lizards: A Model for Brain Evolution'.

[39] D. Robertson, 'Social Control of Sex Reversal in a Coral-reef Fish'

[40] E. Charnov, 'Natural Selection and Sex Change in Pandalid Shrimps: A Test of Life History Theory'.

More challenging to group-selectionism are instances of behaviour or appearance which contradict chromosomal or genetic sexuality, rather than change with it in the cases of transsexualism mentioned here. 'Homosexuality' is a term applied to human behaviour when a person of one sex adopts members of the same sex as love-objects, and this is certainly something which poses a challenge to theories which see sex as benefiting society, the family or the species rather than the individual male or female.

As far as male homosexuality is concerned, we might begin by observing that, in part, confusion about the nature of a male's sexual object is not surprising. As we saw, the low cost of sex cells to males, often combined with low or negligible levels of parental investment in offspring, means that males need not be anything like as selective and discriminating in their choice of objects as females normally are. For instance, male guppies confined in tanks with other males will attempt to mate with each other – an example of apparent ethological 'homosexuality'. In fact, male guppies will mate with any fish that vaguely resembles the female of the species, including females of related species and males of their own who are large enough to suggest the usually larger female guppy. Female choice has, as I mentioned earlier, resulted in male guppies having to 'advertise' their species with their glorious tails, but this is for the benefit of the females of their species, not the males. Guppy 'homosexuality' is therefore nothing more than male sexual indiscriminateness – according to the theory of parental investment, a classically masculine attribute!

If females of some species can behave like males, males can certainly perform like females, with – paradoxically – highly beneficial effects on their role as males. There is an abundance of evidence from ethology that undifferentiated, non-dimorphic males can and do evolve, if not as the dominant form of male, at least as a subordinate, or deviant type.

An example would be the bluegill sunfish. Males are territorial and invite females to lay eggs at their jealously guarded nest sites. However, three kinds of male are found in this species: a large, 'regular', dimorphic male who defends such a territory and courts females; 'little sneakers' who are young males far too small and insignificant to defend territories, but who exploit their small size for amazing turns of speed by lurking around near nests and then darting in at the moment the resident male has succeeded in inducing a female to spawn and instantaneously ejaculating! Finally, sneaks can grow into 'transvestite' males who resemble females externally, but are equipped with testes, instead of ovaries. When a male succeeds in inviting a genuine female to spawn, the transvestite swims up and is willingly accepted as a second mate by the regular male; but instead of laying eggs the transvestite takes advantage of his disguise to fertilize the eggs of the genuine female with his own sperm.[41] In the case of salmon, 'precocious parr', a tiny immature male form, remains in the natal streams, never maturing or going to sea, but waiting to fertilize returning females who do. Again, recent evidence suggests that commercial culling of fully grown males has caused less mature salmon to return to breed, thanks to reduced competition from fully dimorphic males.

Such adaptations are not confined to fish. Red deer populations contain deviant male 'hummels' who never grow antlers and avoid confrontations with regular males, but attempt sneak copulations instead. The same effect is found in stag beetles, where many species include both elaborately horned and unhorned males. Many similar examples exist in scorpion flies, sticklebacks, prairie chicken, elephant seals and Canadian garter-snakes, where female-mimics distract other males from mating with the real females, evidently to the mimics' advantage.

In many species in which 'sneak' fertilization has evolved as a subordinate sexual strategy for males, the males in question are notably less dimorphic than the regular, fully developed males. This comes about either by means of seeming to be female, like the transvestite bluegill sunfish and female-mimic garter-snake, or by being less male in the sense of being less mature, like the precocious salmon parr and the hummel deer. In such cases, the cryptic male uses the appearance of being female or sexually immature to gain the benefit of covert fertilizations, without the costs normally involved in regular male competition. Thus, transvestism and sneak fertilization are deceptive strategies which aim to secure repro-ductive success for their practitioners, not by open competition, but by concealment and deceit. They represent instances of *cryptic sexuality*: disguised sexual behaviour or appearance which exploits the element of camouflage provided by an irregular or non-

---

[41] W. J. Dominey, 'Female Mimicry in Male Bluegill Sunfish: A Genetic Poly-morphism?'.

dimorphic appearance to secure matings by deceit, stealth or surprise.

Here, once again, a valuable new insight is gained by concentrating on the question of the costs and benefits to individuals of sexual behaviour, rather than naively assuming that all such behaviours evolve to benefit the family, group or species. Sexual behaviour and appearances of the kind cited above only seem to be paradoxical if we assume that sexual differentiation ought to be simple, straight-forward and unambiguous because it is aimed at reproducing the group or species. What both classical psychoanalysis and modern biology have found is that this is often not the case but that, on the contrary, 'paradoxical' or seemingly 'unnatural' sexual appearances and behaviour can be both adaptive for the individuals concerned and the natural consequence of seeing the fundamental distinction between the sexes in terms of costs and benefits, not to some social group, but to individual organisms and the genes which they carry.

Sex, subordination and conflict

Such a view as this emphasizes conflict and competition rather than harmony and cooperation between individuals over the question of sexuality. Yet numerous instances exist where organisms exhibit astonishing tolerance of others, and even seem actively to avoid conflict, sometimes to the extent of appearing to act in the interests of the larger group or species, just as the group-selectionist view would suggest.

For instance, in the case of the peacock wrasse males make nests and try to attract females to spawn in them just like bluegill sunfish do. Just like them, some peacock wrasse appear to specialize in 'sneak' fertilization, but not so the largest males. These go in for 'piracy' of nests of other males, who, after a brief conflict, have their nests taken over by *force majeure* and often find that their jealously guarded eggs are being eaten by the intruder.

Yet, surprisingly, males who suffer this fate stay on and look after the fertilized eggs, including those of the pirate who abandons them once females cease to appear! Here, it seems, is a case of 'altruistic' behaviour in the interests, not of the individual nest-building male, but in the greater good of the species as a whole. Why else should the ousted owner lavish care on offspring who are not his? Surely, this is a case of selection operating at the level of the group or species, rather than at that of the individual?

Again, the view advanced above which sees individual advantage and conflict as the key to sex and society can hardly seem to explain the frequent avoidance or ritualization of conflict with others which is widely reported of many animals. It would seem that nature is often nothing like as red in tooth and claw as Darwinism might lead one to expect. For instance, it has been suggested that where animals do not fight to a finish, but break off long before any real harm is done, this is because

> The opponents assess their strength without injury to the weaker. The selection pressure which leads to such ritualized combat is easily understood: It acts as species preserving if the weaker opponent is chased off and excluded from reproduction for the present. However, he should live on so that he may be victorious later.[42]

But consider the evolutionary paradox posed by this way of looking at things. Let us suppose that it were indeed the case and that individuals might possess a gene for ritualized, rather than lethal, aggression because of some benefit it conferred on the species as a whole. Suppose that a mutant appeared who lacked that gene and instead of inhibiting its aggression in the interests of the species, instead carried it through to lethal effect. Obviously, it would have an advantage over its conspecifics who would avoid lethal injury to it even though it inflicted such injuries on them. In practical effect, their restraint would be a form of altruism which would promote the selfish aggression of the mutant and enhance its success in any conflict to the mortal cost of its opponents. Since evolution is driven by differential reproductive success, such a lethal mutation would be advantaged by the existence of the ritualizing gene and would proliferate until all individuals were rid of it.

But quite apart from such purely theoretical difficulties, grave factual objections can be raised to the view that aggression is 'naturally' inhibited for the good of the species. According to another and rather better informed view, 'ethology's conventional wisdom, that animals rarely kill or main each other in intraspecific fighting, has been overstated.'[43]

[42] Eibl-Eibesfeld quoted by Geist, *Mountain Sheep*, p. 231.
[43] Ibid., p. 230.

It is a popular game today, as in the past, to scrutinize the actions of animals and exalt exemplary conduct, thereby justifying it for humans. The fiction that wolves will not kill wolves, that 'dog does not eat dog' is a hapless example of this. Unfortunately for wolves and moralists, the former do kill and consume conspecifics just like other large carnivores – just as we did throughout our history as a species and still do sporadically. It is to the credit of carnivores that they do not indulge in murder wholesale, as is eminently respectable in human societies. Nevertheless, they show an inhibition to kill conspecifics at best within their own social group, much as we do.[44]

Conflict is expensive and dangerous and can result in death, injury or exhaustion which can temporarily or permanently alter an individual's relative standing in this respect:

> The aggressive individual who readily attacks and inflicts painful slashes on his opponent is likely to trigger a desperate defense and emerge injured from his very first fight. Even if victorious he has already compromized his dominance rank, breeding success, health, and life because he is no match for opponents of equal or even smaller size in the many interactions that will follow every day. He is likely to sicken from his wounds, drop in rank, and lose out in the breeding and perhaps succumb to his injuries ... Seen in this light, the reluctance of a dangerously armed dominant to engage with a subordinate is less a consequence of altruism than of self-preservation.[45]

Any animal which comes into conflict with another member of its species may meet an opponent who is equal to itself in fighting ability, is inferior or superior. It will not normally pay an animal to continue a conflict started with an antagonist who is clearly superior, because injuries or exhaustion incurred in that conflict could – and almost certainly would – reduce its chances in later, less unequal ones. Again, individuals who are of superior fighting ability will seldom find it in their self-interest to keep proving the fact to subordinates if the latter avoid conflict with them for the reason just given and if there always remains a chance, however small, that injury or depletion incurred in such contests might prejudice the superior organism's prospects in more critical conflicts.

[44] Ibid., p. 130.
[45] Ibid., pp. 234–5.

Only protagonists who are of roughly equal standing are likely to fight to the point where one or the other has to retire, admit defeat or suffer serious injury or death. Such considerations readily explain why conflict among males is not to be expected on every occasion and in every context, even in a strongly dimorphic species featuring extremes of polygyny. If we take into account the added factor that many animals can both recognize and remember others with whom they have interacted in the past (and perhaps even had an opportunity to spar with in early life), we can see that chronic conflict and permanent antagonism are not to be expected in most cases.

On the contrary, it seems that in general the potential costs to the individual of unequal conflict will cause it to want to conserve its aggression for circumstances where serious fighting is the only way to establish its dominance. This is confirmed in the case of mountain sheep, where

> the most spectacular social interactions of sheep are the dominance fights of rams. They are by no means common, but they may occur at any time of the year when strange rams meet. If the strangers differ conspicuously in horn size, the smaller one at once acts like a subordinate, and a normal dominant–subordinate interaction results between them. If strangers of equal horn size chance upon each other, they have no means to judge each other's fighting potential except by fighting. [46]

In other words, simple self-interest can be seen to reduce the incidence of conflict within a group or species quite markedly once both the costs and the benefits of aggressive behaviour are taken into account. Since the costs will often outweigh any likely benefit, conflict will often be avoided, not for the benefit of the group as such, but for the benefit of the individual concerned.

Cost–benefit calculations of this kind carried out in the currency of parental investment explain the remarkable tolerance of the ousted peacock wrasse mentioned earlier. Although such behaviour may seem to benefit the intruder and therefore to be in his interest, one only has to take the costs and benefits of the situation for the individual nest-builder into account to see that, unsatisfactory as it may be for him as an individual, staying on pays.

[46] Ibid., p. 189.

This is because the resident male has no way of discriminating between the eggs which he has fertilized and those which have been fertilized by the pirate. However, since at least some of the eggs deposited in the nest will have been his, and since he invested very considerable resources in building it in the first place, it will pay him to nurture them along with those of the pirate, rather than abandon them and lose his entire investment. This interpretation is corroborated by the observation that

> takeovers generally occurred after the most active spawning period when nests contained sufficient numbers of eggs to make the nest worth defending by the original male. Only once did a pirate take over a nest with few eggs; here, the former owner abandoned first, followed by the pirate.[47]

Again, the fact that pirates readily ate eggs which they found in a nest at takeover, but only rarely after they had themselves spawned, suggests that for them too their own parental investment in offspring counted for something.

Considerations of this kind also explain much of the apparently 'perverse', 'unnatural' sexual behaviour mentioned earlier. Indeed, apparent 'homosexual' behaviour is often a consequence of the very kinds of conflicts we have been considering. In the case of mountain sheep,

> Subordinate rams mimic the behaviour of an estrous female, and this happens to contain much overt aggression. He literally 'parasitizes' the dominant's inhibition to blast females. . .The inhibition to strike females is there already, and the mimic takes advantage of it.[48]

It seems that 'in sheep society one does not differentiate in one's conduct between males and females, but only between larger (dominant) and smaller (subdominant)' and for this reason 'homosexuality is normal and adaptive for rams.' Dominant rams mount subordinate individuals 'irrespective of the latter's sex and age' because 'only if a sheep can mount another without being punished has it demonstrated dominance.' Furthermore, subordinate males so mounted often respond like females, so that the 'homosexual'

[47] E. Van den Berghe, 'Piracy as an Alternative Tactic for Males'.
[48] Geist, *Mountain Sheep*, p. 235.

behaviour in question has both an active 'masculine' and a passive, 'feminine' partner, despite the fact that both are in reality male.

Although these examples might suggest that rams contest more in the interests of their social standing than in that of their ultimate reproductive success, the fact remains that 'male dominance and breeding success run parallel with horn size, and rams use their horns not only as weapons or shields but also as rank symbols.' Again, not only do 'large-horned rams successfully chase away smaller-horned ones from estrous females and do most of the mounting', but such females 'prefer large-horned males over small-horned males'.[49]

Looked at from this point of view, apparent 'homosexuality' or sex-role reversal only seems paradoxical because it is entirely motivated by the sexual self-interest of the animals in question. Males who act like females in order to avoid damaging conflict with other males or to secure covert matings are just as surely promoting their own ultimate reproductive success as are the more regular males. There is absolutely nothing paradoxical about this as far as evolution is concerned: in either case, whether by regular or irregular male behaviour, males act in such a way that the costs of reproductive competition are minimized while the potential benefits are enhanced.

Although I have used sheep and fish as my principal examples, I could cite similar cases almost endlessly, especially among primates such as chimpanzees. Among the large class of animals who practise copulation, mounting behaviour is practically universally interpreted in the same manner as it is among sheep. In the vast majority of cases mounting connotes superordination and being mounted subordination, so that dominance behaviour, conflict and sexuality become intertwined in a familiar complex which often applies to females just as surely as it does to males.

In general terms we can conclude that, given the high costs involved in contesting dominance with others, cryptic or deceptive strategies might pay, particularly if an animal were notably smaller, more immature or less aggressive than potential antagonists. Yet just as evolutionary self-interest explained avoidance of conflict as well as conflict itself because the latter implies costs as well as benefits, so reproductive self-interest alone can explain apparently

[49] Ibid., pp. 13, 139, 131, 151 and 174.

'perverse', 'unnatural' behaviour, thanks to the benefits which avoiding the costs of 'standard' sexual behaviour can sometimes bring. Not merely sex-role reversal, but bisexuality in all its forms seems an inevitable consequence of the asymmetry of the sexes with regard to parental investment.

As long as the costs and benefits of sexuality fall differentially on the two sexes – and we saw at the beginning that differential costs and benefits are the essence of sexual distinction – it is perhaps inevitable that each should try occasionally to appropriate benefits which normally belong to the other or avoid costs which usually fall predominantly on itself. In a sense it is a case of trying to get the best of both genders by confusing the fundamental divide between the sexes with behaviour, appearance or attributes borrowed from the other.

Evidently, deviation from the normal or predominant sex role can pay individuals in an evolutionary sense even if it seems paradoxical when looked at from the point of view of the presumed benefit of sexual behaviour to the family, group or species. Indeed, sexual pluralism of this kind seems inevitable once individual interests, rather than the ostensible group benefits of sexual behaviour, are seriously taken into account. As almost always seems to be the case, the viewpoint of the individual reveals more than bland generalizations about the group.

The myth of monogamy

The general ideas and terminology outlined above constitute the fundamentals of the theory of parental investment and illustrate its crucial relevance to the understanding of sex in general and to mating strategies, behaviour and evolution in particular. If we now turn to the question of human sexual behaviour, the principal theme of this book, we might begin by considering where, in the system of mating strategies outlined above, human beings belong.

In terms of mating strategies as revealed in forms of human marriage and reported by social scientists, the facts seem to be clear. In a survey of 849 human societies, 708 (or 83 per cent) were habitually or occasionally polygynous (with the split roughly 50/50), 137 (or 16 per cent) were officially monogamous and 4 (less

than 1 per cent) polyandrous.[50] In another survey of 185 contemporary 'primitive' societies, again only 16 per cent were found
to be consistently monogamous.[51] When we note that Western
societies are widely regarded as 'monogamous' despite high rates of
divorce and remarriage and that other, officially 'monogamous'
societies show comparable effects, we might feel even more confident
in concluding that, sociologically speaking, present and recent
human societies have been mainly polygynous. Indeed, even in
officially monogamous societies such as the United States it has
been estimated that between 25,000 and 35,000 polygynous marriages
may exist and surveys reveal that a number of better-off men maintain
two families, each unknown to the other.[52]

The general conclusion that human beings are predominantly
polygynous can be reached despite the perhaps rather misleading
observation that, in human societies overall, the majority of marriages
are of one man to one woman. Some use this as evidence for the
assertion that human beings should be regarded as a predominantly
monogamous species. But this is only possible if we treat 'monogamous' marriages of potential polygynists as equally monogamous
as those of married men in societies which do not practise polygyny.

Yet the fact remains that, given polygyny – that is a few, usually
older and more successful males, each married to a number of
women – and a nearly equitable sex ratio, some men are going to
have no wives, and quite a few will only be able to manage one.
Many of the latter will in due course of time acquire more and even
those who never get beyond one would wish to do so, if given the
chance.

Again, it is worth remembering that one's estimation of the
degree of polygyny is partly determined by whether one counts
husbands or wives. This is because 'the percentage of polygynous
married women will always exceed that of men in a society by at
least 2:1 and sometimes more than 20:1.'[53] By definition, there will
always be more polygynously married women than men in any
polygynous society, and so the estimation of the prevalence of
monogamous marriage is affected by this fact. For instance, in a
recent cross-cultural study of polygyny, I calculated the average

[50] G. P. Murdock, *Ethnographic Atlas*.
[51] C. S. Ford and F. A. Beach, *Patterns of Sexual Behavior*.
[52] Data quoted in D. Symons, *The Evolution of Human Sexuality*, p. 223.
[53] D. R. White, 'Rethinking Polygyny' p. 530.

proportion of polygynously married men in all cases where percentages for both sexes married polygynously were given as amounting to 19 per cent of the total. For polygynously married women the figure was 32 per cent. If we restrict the societies sampled to those where polygyny is 'prevalent and preferred by most men' and ignore those included in the previous figure where it is 'limited to men of higher social class', the average percentage of polygynously married men rises to 44 per cent and that of women to 68 per cent.[54]

Sometimes a society which seems predominantly monogamous if husbands are counted turns out to be polygynous when wives are considered. A case is found in the data for the Miwuyt aborigines of Australia, which we shall be looking at again in a different context later. According to the figures given there,[55] a total of 94 men are monogamously married, compared to a total of 51 polygynously married. On the face of it, this looks like vindicating the view that this group is polygynous to only half the extent that it is monogamous. Nevertheless, figures for women show that whereas again 51 are monogamously married, 156 – or three times as many – are married to polygynists. It seems that what looks like a predominantly monogamous situation from a man's point of view looks completely different from a woman's. To what extent this has anything to do with the conventional estimations of the extent of human monogamy I do not know, but it is a factor worth bearing in mind when one is tempted to dismiss polygyny as 'exceptional', 'unnatural' or limited to minorities.

If, quite apart from the point about divorce and re-marriage made above, we also notice that even in avowedly 'monogamous' societies like our own the facts reveal that a majority of men and a significant proportion of women will have extra-marital affairs at some time or other, the assertion that human beings are 'naturally' monogamous seems to me to be straining a point. In this respect – and bearing in mind the point about bias towards counting only men's marriages mentioned just now – calling human beings 'monogamous' would be like calling species where only a minority of males are mated at any one time 'celibate'. They would not be celibate, but polygynous, in just the same way that human beings are.

[54] Figures and definitions found in White, 'Rethinking Polygyny'.
[55] See below, pp. 122.

If we wish to be completely accurate about it, it might be best to say that human beings are in the main *polygamous* rather than polygynous because, as we saw earlier, this term includes the promiscuity which is nearly everywhere attendant on official monogamy, as well as polygyny and the few rare cases of polyandry. For instance, in the case of the nominally monogamous Mehinaku indians of Amazonia, a study of one village showed that

> the thirty-seven adults were conducting approximately 88 extra-marital affairs ... To put this number in perspective, it would be possible for the villagers to pair off in 340 extramarital (heterosexual) partnerships if they were unrestrainedly promiscuous. If affairs that are in violation of in-law avoidances, the incest taboo, and the respect owed older persons are eliminated, 150 theoretically possible pairings remain. Given that the actual number is 88, I conclude that the villagers' taste for extramarital liaisons is limited primarily by social barriers, such as the incest taboo, and only secondarily by personal preference.[56]

Since this is a society which would probably be labelled 'monogamous' on official, sociological grounds, the actual sexual state of affairs suggests to me that 'polygamous' would be a much more accurate description of the reality, if not of the myth.

Indeed, if we think about it we will readily see that any portrayal of human beings as essentially monogamous must reveal a bias in favour of official, legitimated marriage (or its equivalent) and against covert and illicit affairs. However, there is no such bias latent in the contrary term 'polygamous' which, implicitly at least, recognizes pluralism in sexual partnerships in a manner exactly opposite to that in which 'monogamy' implies sexual exclusiveness. While I can readily see that moralists and those with vested interests in monogamy might want to perpetuate this bias, I can hardly believe that it can be defended if scientific objectivity is our goal. After all, from the basic, biological point of view, sex is sex whether it happens to be licensed by the state or public opinion, within wedlock or outside it.

In other words, unless we are to substitute sociological myth for sexual reality, I think that we have little choice but to conclude that if actual behaviour rather than ideology is our subject, many monogamous societies in fact find their claims in this respect deeply

[56] Gregor, *Anxious Pleasures*, p. 35.

compromised by degrees of promiscuity which, while they may not reach the level seen in the Mehinaku case, nevertheless should make us circumspect. When we take into account the fact that the majority of human societies are officially polygynous, it seems that to portray human beings as essentially monogamous is to take indefensible liberties with the truth.

I believe that confusion on this point is further fostered by misunderstandings of the way in which the sex ratio of males to females is affected by polygyny.[57] In a monogamous species, it is easy to understand why the sex ratio should be one-to-one because each female is ideally mated to each male. We know that the human sex ratio (at least at reproductive age) is roughly one-to-one, and so this seems to suggest monogamy as a natural outcome. In the words of Dr Arbuthnot, 'Polygamy is contrary to the Law of Nature and Justice, and to the Propagation of the Human Race; for where Males and Females are in equal number, if one Man takes Twenty Wives, Nineteen Men must live in Celibacy, which is repugnant to the Design of Nature; nor is it probable that Twenty Women will be so well impregnated by one Man as by Twenty.'[58]

Nevertheless, and ignoring for a moment the question of how well impregnated a number of women may be by one man, the fact remains that even under polygynous conditions where only a minority of males actually mates, an equitable sex ratio is the most likely outcome.[59] This occurs despite the fact that it seems wasteful and contrary to the interests of the species. Indeed, if evolution did in fact select for traits which benefited the species as a whole one would definitely expect a sex ratio which reflected the degree of polygyny, with an excess of females over males corresponding to the ratio of breeding males to females. In a species where, for instance, on average ten females were mated to each male, a 10:1 sex ratio in favour of females would be expected.

In reality, this almost never occurs, ultimately for the reason that evolution does not select for factors which benefit the species as a whole but rather for those which confer reproductive success on its individual members. In the hypothetical case we have just considered, even though only one male in ten may actually mate, any male that does do so is ten times more successful than any that does not. In

[57] For instance, see A. Santangelo, *Il Giardino dell'Eden*, p. 96.
[58] Quoted by J. Hartung, 'Polygyny and Inheritance of Wealth'.
[59] R. A. Fisher, *The Genetical Theory of Natural Selection*, pp. 142–3.

terms of parental investment (and assuming random mating success), males who do mate are ten times more valuable to their parents. Consequently, equal numbers of males and females are to be expected because the potential failure of any one male to mate is exactly compensated for by the potential success of another. All other things being equal, natural selection should hold the sex ratio at reproductive age near to unity in both monogamous and polygynous species, and it usually does.

In terms of an individual's, as opposed to a species' reproductive success, it is clear that parents who produced more females than males in a polygynous situation such as that outlined above would find that they were conferring an advantage on other parents who produced more males. This is because the reproductive success of the other members of the population would be increased by the added number of females available to each of their males that did mate and, assuming that individuals could not know reliably in advance which males would succeed in mating, producing excess females would be tantamount to 'altruistically' adjusting the sex ratio for the benefit of the species. Effectively, such 'altruism' would be self-defeating because it would reward 'non-altruists' who, for instance, switched to producing more males to exploit the greater supply of available females.

Exactly the same would occur if individuals produced an excess of males. This is because, all other things being equal (and assuming again that no individual producer of males could know which males would ultimately be most successful in mating), excess production of males would merely reduce the likelihood of any particular male being successful by exactly the same ratio as the number of males to females was increased.

For instance, imagine that in a small population where only one in ten males mates, some parents could produce enough males to cause a 10 per cent swing in the sex ratio in their favour. One of two things could happen, depending on what independent factors control the ratio of mating males to females. Assuming that all other things are held equal, the ratio of mating males to females will either stay the same, or it will improve, because of the extra supply of males. In the first case only one in eleven males will now mate, thanks to the fact that there are now 10 per cent more males, of whom only 10 per cent can actually mate. As far as the parents providing the extra males are concerned, their efforts have merely reduced the reproductive success of males in exactly the same proportion as they have increased the total number, namely, by 10 per cent. In other words, the quantitative increase in males is cancelled out by a qualitative decline in the value of the investment in the additional males.

In the second case, where the 10 per cent extra males succeed in increasing the proportion of breeding males (shall we say, from one-in-ten to one-in-nine), the fact that the total number of breeding females is the same means that a 10 per cent increase in the number of males has reduced each male's total holding of females by the same amount. Since the reproductive success of males is directly proportional to the number of females each male has, the net gain to parents investing in extra males is again zero, thanks to the fact that, although more males are mating, each one is mated to a number of females exactly discounted by the increase in the total number of mating males. This time a qualitative increase in terms of improved mating success of the additional males is cancelled out by a quantitative decline in their reproductive success represented by the fact that additional males are mated to fewer females than would have been the case had they not existed.

In either case, the increase in males would also have the effect of raising the reproductive success per unit of parental investment in females relative to such males, making increased investment in females preferable (because we assume that all females mate). In other words, the benefit conferred on other parents' females by producing more males would exactly cancel out the apparent benefit conferred on one's own offspring by making more of them males. The net result is that, in species where both parents contribute the same number of chromosomes to the offspring and all other things are equal (and we shall see later that they are by no means always so) it will pay individual parents not to invest disproportionately in females rather than males but to aim for an equitable sex ratio.

Consequently, it cannot be assumed that each individual male is predestined to be mated to each individual female merely by virtue of the fact that the sex ratio of sexually mature males to females is close to unity. On the contrary, what the unitary sex ratio actually indicates is that selection economizes on the reproductive success of individual parents' investment in their offspring, not on the number of mating males and females in the total population. Indeed, the fact that it demonstrably does not do so is one of the most powerful arguments against the case for selection operating at the level of the family, group or species.

If we now return to our consideration of human mating strategies and recall my opening remarks about the fundamental importance of primal hunting and gathering societies and the relative adaptive insignificance of recent, post-Neolithic human existence, we must first dispose of another widely held, but unsupportable prejudice. For what are probably the best possible reasons (related to the quality of the research done, the eminence of the researchers and the institutions from which they came, etc.) there has grown up recently a widespread prejudice which regards present-day African hunter–gatherers as paradigmatic examples of the type. According to one recent textbook, for instance, the unpronounceable !Kung Bushmen (the '!' represents a tongue click) 'are not just another tribe ... They are as near as we can get to a pristine society living as all humankind lived for most of our evolutionary history.'[60]

So widely accepted – and indeed axiomatic – has this view become that any doubts about it tend to be regarded as heresy; but, nevertheless, one or two major problems do exist. The first is the realization that the !Kung live in an environment almost certainly quite different from that of primeval African hunters and gatherers – and, furthermore, one featuring an unusually high dependence on vegetation, rather than meat. This is because although desert and semi-arid regions such as the Kalahari are not well stocked with game animals, they do contain numerous moisture-preserving plants which constitute a major all-year-round food resource for hunter–gatherers such as the !Kung.

By contrast, the better watered savanna grasslands of Africa provide much less in the way of a readily accessible vegetable diet for human beings, but much more in the way of game. It is precisely in these richly stocked grasslands that hominid hunting seems to have begun and the availability of large game animals suggests that reliance on meat was originally much greater than it has come to be among present-day African hunter–gatherers. Driven into marginal desert and forest habitats by pastoralists who now exploit the grasslands for their domesticated herds, peoples like the !Kung, Mbuti or

[60] Daly and Wilson, *Sex, Evolution and Behavior*, p. 127.

Hadza can hardly be regarded as fully representative of the first, savanna hunter–gatherers.[61]

The second problem involved in regarding present-day African hunter–gatherers as paradigmatic of the type is that such a view totally ignores the much greater numbers of hunter–gatherers in Australia who can genuinely claim to be direct descendants of pre-Neolithic human cultures still living – at least until very recently – in their original habitat. The situation is exactly like that in which naturalists took only the marsupials of South America as representative of the type, or that in which sociologists who discussed industrial societies habitually cited Sweden and seldom if ever took note of the USA or Japan. The contrast between Australian and South American marsupials or the disparity in populations and importance of the cultures concerned in the sociological example are directly comparable to that between the African hunter–gatherers so compulsively cited in the modern literature and the aboriginal peoples of Australia.

The Australian aborigines, unlike African hunter–gatherers, had no contact with post-Neolithic cultures until the arrival of Europeans in recent times; unlike the former they were not driven into marginal habitats by migrations of agriculturalists and pastoralists and did not on occasions come to share their languages or to trade with them. They, unlike recent hunter–gatherers in Africa, numbered at least a quarter of a million at the time of first contact with Europeans and featured over 500 linguistically distinct tribal groups. Of these, the more southern lacked even the bow and arrow; and all Australian aborigines gave evidence of having remained at the hunter–gatherer stage of social evolution since the land bridges with Asia were closed some time before the Neolithic Revolution was to occur elsewhere.

Consequently, a reliance on the fashionable African data on hunters and gatherers seems to me to be selective, to say the least. I propose largely to ignore them on the grounds that, having come into contact with more advanced cultures, these peoples cannot be relied upon to provide any kind of picture of human cultural or sociological adaptations prior to the coming of agriculture. The total isolation of the Australian aborigines, at least until the arrival of Europeans, however, seems to me to make them as valid for comparison with primal, pre-agricultural hunter and gatherer so-

---

[61] R. Foley, 'A Reconsideration of the Role of Predation on Large Mammals in Tropical Hunter–gatherer Adaptation'.

cieties as that of the marsupials of Australia makes them valid evidence of zoological adaptations immediately prior to the radiation of the eutherian mammals.

The fact that all Australian aboriginal societies were polygynous, some extremely so, suggests that the trend outlined by the world-wide comparative data is reliable and that, sociologically speaking, human societies are predominantly polygynous and in hunter-gatherer prehistory were probably almost exclusively so.

If we now turn to the evidence from human physical adaptations the predominantly polygynous picture is reinforced by a number of particular pieces of evidence.

First, *sexual dimorphism*: males show a 5 to 12 per cent excess of height over females in all societies, and in traditionally polygynous ones this tendency is even more pronounced.[62] Again, despite the fact that on average women have more than a fourth of their body weight in the form of fat compared to roughly half that figure in the case of men, they still weigh on average about 20 per cent less.[63] Quite apart from such quantitative disparities, the qualitative factor seems to be independently dimorphic in the sense that female fat-content is presumably related to provision for possible pregnancies (which probably explains why ovulation ceases if weight drops below a critical value), whereas male muscle development is related to inter-male competition and conflict.

Along with these overall differences in size, weight and constitution, we must note secondary sexual characteristics which are distinctive of the sexes such as deepened voices, beards, baldness and broader shoulders in men; breasts, broader hips, rounded figure and more youthful complexion in women. Together all these factors add up to a significant and unmistakable degree of sexual dimorphism (see figure 1).

Inevitably, this is reflected in sexual preferences. For example, in the case of the Mehinaku,

> a youthful woman with long sleek hair, heavy yet firm calves and thighs, large breasts and nipples, small, close-set eyes, little body hair and 'attractive genitals' is an avidly sought after sexual partner.

[62] R. D. Alexander et al., 'Sexual Dimorphism and Breeding Systems in Pinnipeds, Ungulates, Primates and Humans'.

[63] J. H. Crook, 'Sexual Selection, Dimorphism and Social Organization in the Primates'.

CHIMPANZEE   GORILLA   HUMAN

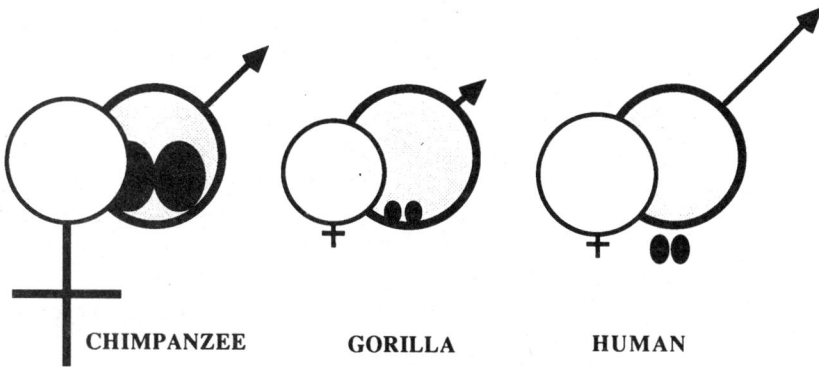

Large circles indicate approximate body size of sexes;
+ Indicates relative largest size of female genitals;
↗ Indicates relative size of penis in erection;
•• Indicates approximate relative size of testes.

Figure 1 Sexual dimorphism, penis and testis size in three primate species
Source: Modified and redrawn from R. V. Short, 'Sexual Selection and Its Component Parts: Somatic and Genital Selection as Illustrated by Man and the Great Apes', *Advances in the Study of Behavior,* 9 (1979)

Appearance in men is also important. A heavily muscled, imposingly built man is likely to accumulate many girlfriends, while a small man ... fares badly. The mere fact of height creates a measurable advantage. Men over 5'4" had an average of six girlfriends at the time of my study, while those under 5'4" had only 3.4 girlfriends.[64]

Although the author adds the qualification that 'as in our own society, men who are socially successful are more attractive to women as sexual partners,' he also points out that

on the average, tall men are more likely than short men to sponsor rituals, be wealthy, have many girlfriends, and become village chiefs. It is they, to the near exclusion of the short, who monopolize the positions of power and influence that are inter-twined with the concept of masculinity. Height prejudice denies the short man an equal chance at realizing his manhood. [65]

[64] Gregor, *Anxious Pleasures,* pp. 35–6.
[65] Ibid., p. 144.

Admittedly, in some species sexual dimorphism can reflect differences in diet and habitat between the sexes, but in the human case
the fact that such differences would presumably be related to male
hunting need not mean that they are irrelevant to male competition
for females. On the contrary, if we concede that greater male
muscularity of build, height, etc. are hunting adaptations, then the
fact that females were presumably reliant on such males for part of
their diet might have meant that female choice of successful hunters
took over the same kind of selective role as crude inter-male hostility
(a conclusion reinforced by the observation that, among modern
hunter–gatherers, hunting success is highly esteemed by women
and the bestowers of women).

If we recall the fact that sexual dimorphism is normally totally
absent or much reduced in habitually monogamous species, the
degree of sexual dimorphism found in modern human beings is
hardly likely to be indicative of an evolutionary past characterized
by monogamy.

A second piece of evidence which points to the same kind of
conclusion is *sexual bimaturism*: the fact that females of our species
become sexually mature notably earlier than do males. Thus,
although men normally mature as somewhat taller than women,
girls between twelve and fourteen years of age in England nevertheless have an average height in excess of the average for males of
the same age and gain height fastest at about twelve years of age,
whereas boys do so at approximately fifteen. On average, girls reach
their full height soon after age fifteen, whereas boys do not reach
theirs until about eighteen.[66] In general, this kind of discrepancy is
less likely in a monogamous species than in a polygamous one,
where later male maturation is related to the advantages of delaying
maturity in order to acquire male attributes (such as size) needed
for breeding success.

Again, *differential life-expectancy* with regard to the two sexes is also
evidence of polygyny, since what kills males off faster than females
are the direct and indirect results of their specific male adaptations.
In human males for instance, the 5 per cent elevation of resting
metabolic rate believed to be induced by the male sex hormone
testosterone seems to contribute to shortened life-expectancy. So

[66] Tanner–Whitehouse–Takaishi data, Institute of Child Health, London, and R.
V. Short, 'Oestrous and Menstrual Cycles', in C. R. Austin and R. V. Short (eds),
Reproduction in Mammals, Book 3: *Hormonal Control of Reproduction*, pp. 137–41.

does the greater aggressiveness towards other males which testosterone in particular, and male adaptations in general, seem to produce. The facts that violent and serious property-related crime are virtual male monopolies and that 'homicidal conflict is overwhelmingly a male affair ... in *all* societies'[67] suggest that aggressiveness carries serious costs, as well as benefits. In modern societies, as in nearly all others, 'males die faster than females, at all ages and for all segments of the population and regardless of the level of mortality.' [68]

The greater vulnerability of males is independent of social and most environmental factors, for it is even more marked *before birth* where the sex ratio for fetuses at three months may be as much as 120:100 males to females, falling to 106:100 at birth. The fact that it exceeds a one-to-one ratio at birth reveals the difference in mortality that intervenes before reproductive age is achieved. Again, the fact that males also die sooner than females in many other species points to biological rather than social causes and this conclusion is reinforced by the observation that 'in some human populations the smallest difference between the sexes in mortality rates occurs when sex-role differentiation is the greatest, that is, in early adulthood.'[69] Most diseases, infections, accidents, homicides, suicides and traumas kill males preferentially.

This is all in sharp and dramatic contrast to most monogamous birds, for instance, where females suffer slightly higher mortality than males – probably thanks to greatly reduced inter-male conflict but enhanced female parental investment, exacting a differential cost in female survival.[70] Again, if human beings were essentially monogamous, as the majority of birds are, one would wonder why they do not show a comparable tendency. The fact that they show exactly the opposite tendency, despite the high costs of parental investment to human mothers represented by pregnancy, lactation and extended child-care, suggests that the human case is not comparable.

An alternative explanation for male vulnerability might be derived from the fact that the male, unlike the female, has an 'unguarded' or unpaired X chromosome. Chromosomes are bundles of genes,

67 Daly and Wilson, *Sex, Evolution and Behavior*, p. 301.
68 Trivers, *Social Evolution*, p. 302.
69 Ibid., p. 306.
70 Ibid., pp. 312–13.

which are the basic units of heredity now known to control the development of an organism. In many cases, such as human beings, they come in pairs. The X chromosome is paired with another X in females, but with the Y chromosome in males. It is this Y chromosome which carries the gene for masculinity. This is significant because the fact that human beings, like most sexually reproducing organisms, have two complete sets of chromosomes allows for the possibility of the inheritance of so-called *recessive* genes.

A recessive gene is one which might, for instance, cause a disability like haemophilia, but whose unfortunate influence is not expressed in the individual who possesses it if the corresponding gene on the paired chromosome is *dominant* and suppresses it. Since the X chromosome in males is not paired with another X as it is in females, the possibility arises that this 'unguarded' chromosome might be the cause of differential male mortality.

As Robert Trivers concludes in his excellent summary of the available evidence, the trouble with this explanation is that in fish, for instance, both sexes are XX and yet males still show higher mortality. Again, in birds we find the exact opposite of the situation in mammals: males are XX and females XY (sometimes rendered ZZ and ZW), yet, other things being equal, males still show higher mortality rates. Finally, studies of castrated males in a range of species, including human beings, shows that it is the presence of male sex hormones and not chromosomal constitution which shortens life-expectancy; a castrated man can expect to live on average about as long as a woman.

Summarizing the evidence, Robert Trivers concludes:

> It is easy to show that differential male mortality in humans does not result from a difference in the way the two sexes are treated by the larger society. It is likewise clear that the difference in sex chromosomes between the two sexes cannot account for the pattern of differential male mortality. Instead it seems likely that males suffer higher mortality than do females because in the past they have enjoyed higher potential reproductive success, and this has selected for traits that are positively associated with high reproductive success but at a cost of decreased survival. Evidence ... suggests that males suffer differential mortality as a direct or indirect consequences of male–male competition for females.[71]

[71] Ibid., p. 314.

If, as we saw earlier, even minor sexual dimorphism can indicate a significant element of polygamous behaviour in an otherwise monogamous species of swallow, it seems that the very notable sexual dimorphism, bimaturism and differential life-expectancy found in human beings indicates an even greater degree of polygamy. These facts, along with the comparative sociological findings and the evidence of primal hunter–gatherers in Australia, suggest that monogamy among human beings may be a comparatively modern development and, as an evolutionary adaptation, something of a myth.

The multi-male mistake

Having dealt with the pro-African hunter–gatherer prejudice, we must now notice another fallacy in modern thinking about human beings and their origins which is just as misleading. It concerns an issue which touches on a further question regarding human mating which has remained unresolved so far.

Putting purely sociological data to one side for a moment, we have already seen that human beings, at least on the evidence of their evolutionary adaptations, do not seem to be predisposed to monogamy to anything like the extent that many birds, for instance, evidently are. Nevertheless, we might still wish to know to what kind of polygamy our species tends because, as we noticed earlier, three possibilities exist: *promiscuity* (many males mated to many females), *polyandry* (many males mated to one female) or *polygyny* (many females mated to one male).

To the best of my knowledge, polyandry has never been seriously suggested as a primal sociological adaptation for our hominid ancestors. However, to the extent that polyandry can be seen as a multi-male breeding system, it is not too far from another multi-male model which has been extremely fashionable among those interested in reconstructing hominid social origins. The fashion to which I refer is that which regards modern chimpanzees in general, and the largely promiscuous kind of society depicted in some modern accounts of them in particular, as indicative of human evolutionary origins.

On the question of shared evolutionary ancestors, there is absolutely no doubt that man and chimpanzee are more closely related than man is to any other species. Recent biochemical tests have proved this conclusively, if any further proof were needed. But the fact that modern human beings and present-day chimpanzees shared a common ancestor in the relatively recent evolutionary past does not necessarily mean that present chimpanzee adaptations, life-style or social structure are indicative of human origins. Modern lions and tigers may share a common ancestor in a similar kind of way, but this does not mean that tigers, for instance, should be expected to show a social structure or pattern of behaviour comparable to those of lions. On the contrary, we find that one has adapted to a life as a solitary hunter in jungles, whereas the other is adapted to a social existence on the open savanna, with all the implications that has for behaviour and social organization.

Despite widespread popularization of the view that because human beings and modern chimpanzees share a common evolutionary ancestor they must also share common social adaptations, numerous important pieces of evidence suggest that, whereas the first statement is true, the second cannot be. The critical piece of evidence relates to the relative size of the testes in human beings and chimpanzees.

I suspect that part of the appeal of the hominids-were-like-modern-chimps theory lies in the appealing accounts which a number of popular writers have given of an easy-going, affable, non-competitive life-style which chimpanzees are supposed to have, particularly with regard to sex. Although more objective accounts show that things are not quite so simple, it remains a fact that the life-style of the chimpanzee produces groups of fluctuating size and composition in which a number of males copulate quite frequently with a number of females. Since individual males cannot monopolize groups of females for any extended period of time, a promiscuous and relatively non-competitive breeding pattern emerges – which accounts for the mild sexual dimorphism in the species.[72]

It also accounts for the relatively large size of the testes in chimpanzees, since 'primate species with multimale breeding systems consistently have larger testes than those with unimale breeding groups.'[73] This is because, lacking an ability to monopolize females for copulation when they are most likely to conceive (the period

[72] For a recent summary see T. Nishida and M. Hiraiwa-Hasegawa, 'Chimpanzees and Bonobos: Cooperative Relationships among Males'.
[73] R. D. Martin and R. M. May, 'Outward Signs of Breeding'.

known as *estrus*), male chimpanzees instead have to compete as inseminators: the more sperm any one particular male deposits in any one particular female being serviced at the same time by other males, the greater the chance that the resulting offspring will be his, and not the others'. Since testis size is directly correlated with the quantity of sperm produced, such 'sperm competition' will select for large testes in multi-male breeders.

As might be expected, this is a generalization which applies more widely. Another myth comparable to that which depicts chimpanzee males as sexually cooperative is the one which holds that grey whales need 'helper' males for successful copulation, evidently to 'support' or 'wedge' the female in the mating position. If this were true it would pose serious problems for evolutionary theory because it would be very hard to show how the 'helping' behaviour could be selected since, by definition, it confers reproductive success on *another* male.

Fortunately for evolutionary theory, recent research suggests that large testis size in grey whales indicates that males compete with one another inside the female's reproductive tract by means of sperm competition of exactly the same kind as is found in chimpanzees. Presumably this is because the life-style and environment of grey whales, like the otherwise very different life-style and habitat of chimpanzees, impose constraints which makes competition among males in any other form less worthwhile. In short, so-called 'helpers' are not helping so much as waiting for a chance to copulate themselves and let their sperm carry on the conflict with other males which they otherwise avoid. Here, as so often seems to be the case, apparent contradictions of basic evolutionary principles turn out to be errors of fact, not refutations of the theory. [74]

If we now look for comparable evidence of sperm competition, with its implication of a multi-male breeding system analogous to that found in chimpanzees, in modern men we will be disappointed. Although the human penis is, relatively and absolutely, far larger than that of the chimpanzee or gorilla, the testes are far smaller in relation to the rest of the body than those of the chimpanzee and fall into the range of those primates who exhibit breeding groups with single, rather than multiple, males. This evidence appears conclusively to rule out of account the popular idea of the prom-

---

[74] K. Ralls and R. Brownwell Jr, 'Sperm Competition in Grey Whales'.

iscuous multi-male breeding.group as a likely precursor of modern human social structures (see figure 1 on p. 57).[75]

However, testis size alone cannot tell us whether the one-male group concerned is likely to be monogamous or polygynous, only that it is extremely unlikely that (barring dramatic reduction in the size of testes in very recent evolution) modern human beings are descended from ancestors who lived in multi-male groups. In order to find evidence which can determine whether monogamous one-male or polygynous one-male breeding systems were the case among ancestral humans we need to look at other factors which, along with that just reviewed, cast further doubt on the theory which regards chimpanzee society, or some monogamous variant of it, as the evolutionary model for human beings.

The fact that human beings and chimpanzees share a common ancestor does not preclude the possibility that either or both species could have differentiated very considerably with regard to their characteristic mating systems and sexual strategies. Nor does it rule out the possibility that one or other or both might show evidence of what we might term *convergent evolution* or *evolutionary parallelism* with some other species. In the case of modern human beings there is some impressive evidence that this has indeed occurred and that, since parting company with the common ancestor which we share with modern chimpanzees, hominid evolution has been influenced by factors which have also shaped the evolution of another primate, albeit it one not particularly closely related to us from the genetic point of view.

The list of these differences is long and subtle, but the most obvious are features like the human hand, which is unlike that of the chimpanzee in its capacity for bringing the thumb into opposition with the relatively short and agile fingers. Again, the notable reduction in the size of the canine teeth in humans would appear to call for some explanation, especially since we are a species marked by moderate sexual dimorphism and since large canine teeth are a characteristic of chimpanzee males.

Teeth and hand are, of course, especially indicative because so obviously responsive to diet and habitat, but other features also call for comment. Both human males and females differ from chim-

[75] A. H. Harcourt et al., 'Testis Weight, Body Weight and Breeding System in Primates'.

panzees in possessing secondary sexual displays on the upper front of the body – breasts in the case of women, beards in the case of men. Both, again unlike chimpanzees, possess luxuriant head hair which grows down to form a cape over the shoulders. Both have buttocks and women, unlike chimpanzee females, do not advertise ovulation by a prominent sexual swelling. Whilst these last-mentioned differences may not be surprising in the light of Darwin's observation that 'characters gained through sexual selection often differ in closely-related forms to an extraordinary degree',[76] the fact remains that, with the exception of the final one, they nevertheless appear to resemble the adaptations of a primate only distantly related to human beings, the gelada baboon.[77]

Geladas have testes which, relative to body weight, are of closely comparable size to those of modern men.[78] They also show evidence of considerable sexual dimorphism quite apart from the secondary sexual features already mentioned. The gelada is the most terrestrial of all primates – mankind only excepted – and is characterized sociologically by marked *polygyny*. This, like all the other adaptations mentioned above, is largely determined by life-style and habitat. Unlike chimpanzees, which inhabit woodland and subsist preferentially on fruit, geladas are found today on dry savanna grasslands eating a diet which consists of grass blades, seeds, tubers and berries. Since the carrying capacity of the land is low and since food comes in such a small and highly dispersed form, only small groups can subsist and must continually wander over a considerable range. Single dominant males can and therefore do monopolize groups of females numbering up to about ten and averaging four. Other sexually mature males are driven out of the one-male group by the dominant male and associate with other unmated males in 'all-male groups'.[79]

The canine teeth of the gelada are relatively small, despite its otherwise notable sexual dimorphism, enabling it to chew the hard and numerous particles of the diet with a rotary chewing action like

[76] Darwin, *The Descent of Man*, p. 150.

[77] C. Jolly, 'The Seed-eaters: A New Model of Hominid Differentiation Based on a Baboon Analogy'.

[78] Harcourt et al., 'Testis Weight'. Percentage ratios of body-to-testis weight given here are: gelada baboon, 0.08; man, 0.06; chimpanzee (Pan troglodytes), 0.27.

[79] Nevertheless, followers, subordinate males usually younger or older than the dominant, are also commonly found in these groups. R. Dunbar, *Reproductive Decisions*, pp. 17–18.

that with which we might eat peanuts or some other comparable food which comes in the form of numerous granular particles. Large canines need a gap in the lower dentition to accommodate them but then lock the jaw when it is closed, making such a rotary action difficult or impossible. Geladas, like human beings, have a high, arched palate and a large, labile tongue with which food can be returned between the grinding surfaces of the molar teeth, which also show notable similarities to those of human beings.

The ability to bring the thumb and index finger into opposition in both humans and geladas is notably good by contrast to the near genetic relatives of both species. Presumably this originates in geladas needing to be able to pick up small food items, much as humans might pick up a pin. A secondary sexual display on the chest of the female suggests, along with possession of rudimentary buttocks, that it is their characteristic sitting posture while feeding which means that a perineal display, used by most other female primates (such as chimpanzees), would not be as noticeable as evolution would have it be and that the display in question has therefore been transferred elsewhere.

All in all, it seems not unlikely that, far from originating in a chimpanzee-style society, modern human beings are descended from hominid ancestors whose way of life resembled that of the modern gelada baboon and probably featured a comparable, or very similar, dietary specialization. If these notable adaptations are indeed evidence of parallel or convergent evolution between the two species – and it is important to recall that it is these features which also distinguish them both from their near evolutionary relatives – then the implication must be that it is not to chimpanzees, but to baboons that we should look for the breeding system characteristic of our species.

But however that may be, it is by no means unreasonable to conclude that whatever view we take should be firmly founded on the theory of parental investment and should consider both the costs and the benefits of sexual behaviour to the individuals concerned. In adopting such a dynamic and individualistic view of the question of sexuality, both its evolution and – at least in the human case – its relation with dynamic psychology are likely to be increasingly relevant. It is to a more detailed consideration of these questions that we must now turn.

2

Oedipal Reproductive
Strategies

Parent–offspring conflict from the child's point of view

As we have seen, the theory of parental investment yields fresh insights for our understanding of sex throughout nature. Specifically, this comes about by looking at sexual behaviour in terms of costs and benefits to individuals, rather than in the traditional terms of some putative benefit to the family, group or species. But we need not stop there. Clearly, implicit in the idea is a comparable cost–benefit analysis applied not just to individuals of each sex in adulthood, but to the individual offspring too, in its relationships both with other offspring and with its parents.

If it is indeed the case that anthropologists and sociologists seldom took much notice of individuals as such, then it is most certainly true that they hardly ever spoke to children and that the question of the child's interests in social groups such as the family was almost never considered. This was because the predominant 'cultural-determinist' school of thought in sociology and anthropology assumed that children were the largely passive victims of *socialization*, understood as a process whereby cultural values are instilled into the young.

Academic psychology was just as bad. The doctrine of *behaviourism* taught that a human being, like any other organism, could be conditioned to perform more or less any action by appropriate punishment, reward and reinforcement. The child in particular was seen like Pavlov's captive dogs – conditionable by stimulus and reinforcement, but largely passive in the process. Although psychology emphasized individual conditioning and sociology its collective equivalent, both came to much the same thing as far as the child was concerned. Either way, the child counted for little, the conditioning influence for practically everything.

But sociobiology, like psychoanalysis, did take account of the interests of offspring. Psychoanalysis did so implicitly from the beginning, because of its dynamic model of human psychology, and directly once modern child-analysis became possible. Sociobiology did so in its turn, largely thanks to the insights provided by the theory of parental investment and its application to the issue of *parent–offspring conflict*.

Just as the theory of parental investment followed in the footsteps of psychoanalysis by looking at sex from the point of view of both the costs and the benefits to the individuals concerned rather than in terms of some supposed benefit to the group, society or species, so it retraced the path of Freud in looking at parental care itself from the point of view of both parent *and* offspring. As Robert Trivers pointed out in a classic paper which initiated this advance, 'once one imagines offspring as *actors* in this interaction, then conflict must be assumed to lie at the heart of sexual reproduction itself – an offspring attempting from the very beginning to maximize its reproductive success would presumably want more investment than the parent is selected to give.'[1]

An example might be crying and smiling in human infants. As Trivers points out in his paper, an offspring cannot physically dominate its parents, but it can attempt to manipulate them by purely psychological means. For instance, it could cry when it wanted something like food or attention and smile when it was gratified. Evidently, human infants do both of these things, and one can well imagine socialization theorists and behaviourists agreeing with Trivers when he comments that both parent and offspring can be seen to benefit from this system of communication.

However, theorists of neither persuasion would go on to point out the obvious conclusion: 'But once such a system has evolved, the offspring can begin to employ it out of context. The offspring can cry not only when it is famished but also when it merely wants more food than the parent is selected to give. Likewise, it can begin to withhold its smile until it has gotten its way.'[2] Although behaviourists and sociologists might view such 'inappropriate' behaviour as crying when not really hungry and not smiling despite being physically

[1] R. Trivers, 'Parent–Offspring Conflict'.
[2] Ibid., p. 257.

satisfied as 'deviant', 'pathological' or the outcome of some error in conditioning and reinforcement, from the child's neglected and despised point of view it is entirely normal, and completely natural. So natural is it that if infants amplify their distress signals to the limit to get what they can, parents correspondingly re-calibrate their responses so that, for instance, they respond much less to cries of distress in a young child than they do to those in an older one or an adult. Common observation shows that adults will pass screaming children in the street without so much as a glance, but pay close attention to an adult who is quietly sobbing or showing even more subtle signs of distress (albeit without wanting to get involved).

Once we allow ourselves to entertain the possibility that, despite its lack of experience, small physical size and social subordination, the infant is in fact a significant participant in the interaction, rather than a mere passive recipient of parental care, it becomes evident that both parents and offspring are active protagonists and both interact dynamically on a more or less equal footing. In the view of both socialization theory and psychological behaviourism this was not so. By ignoring the independent interest of the offspring and its ability to respond dynamically to the actions of the parent these theories implicitly took the parental interest as the only one that mattered, assuming with what now seems astonishing naivete that parental and offspring interest were one and the same. Effectively, this is exactly the same situation which we noticed in the first essay with regard to sex: cost and benefit to individuals were traditionally ignored in behavioural science in favour of some presumed common benefit to all involved.

But, just as we saw that new and surprising insights result if this complacent view of sex is abandoned for a more dynamic one which does take account of differences of interest in the parties concerned, so parent–offspring relations can be seen in a completely new light if a similar change of perspective is adopted. Furthermore, just as we saw that the new view of sexuality in biology had much in common with the Freudian one, so we shall find even greater coincidences of insight in the case of parent–child relations which in psychoanalysis were from the beginning seen from a dynamic, objective viewpoint.[3]

---

[3] For further comment and substantiation on this point see Badcock, *Essential Freud*.

The riddle of regression

One example of the convergence of sociobiological and psycho-analytic insights might be Robert Trivers's suggestion regarding the possible evolutionary basis of *regression*.[4] In species in which parental care involves active provisioning or protection of the young by one or both parents, investment in them is likely to be maximized early on and to decline with time, gradually or sharply as the case may be. Either way, younger offspring usually demand and get more investment than older, more mature ones. But this means that any particular offspring might be able to secure more investment from a parent if it could mislead the parent about its exact level of development. In other words, an offspring which acted or appeared 'younger' than it really was might secure more in the way of parental care, provisioning, protection, or whatever.

As a behavioural adaptation, regression would appear to have a sound evolutionary basis, at least in the young. But regression is a factor long known to psychoanalysts and a central tenet of Freudian psychology. In the past it has been described from the three classical psychoanalytic viewpoints mentioned in the introduction: psychological *dynamics* (where it is seen as an active desire to return to the past), *topography* or *structure* (where regression normally involves 'lower', more primitive or repressed elements) and *quantitative* factors (where it is often a question of past 'fixations' attracting the retrogressive desire). The development implied by Trivers's observation would add a fourth, evolutionary and adaptive description with the benefits suggested in my introductory discussion.

But why should regression be a factor in *adult* psychology? Surely Trivers's idea cannot apply here. How could regressive behaviour pay adults, who, far from having parents to invest in them, might actually be parents themselves? Although almost obvious and not very controversial in the case of children, regression as a real psychological factor in adults seems paradoxical, to say the least.

Yet, for all that, any psychoanalyst would have to insist that, paradoxical and apparently inexplicable or not, regression is a very commonly encountered factor in adults, as well as children. Indeed, one of the major new insights contained in Freud's pioneering *Three Essays* was the observation that the unmistakably childish

---

[4] Trivers, 'Parent–Offspring Conflict', p. 257.

element in many of the sexual perversions of adults could only be explained by regression. Examples abound from pleasure in sucking, kissing and licking (evidently a regression to the oral stage of infantile sexuality), caning in particular and sado-masochism in general (the same, but to the anal stage) and masturbation, exhibition-ism and voyeurism (which are found to be related to the third, phallic stage).[5]

On this occasion, it seems that the addition of a fourth, evolu-tionary explanation alongside the traditional psychoanalytic ones has done us no good: we are still lacking any explanation of how, from an evolutionary point of view, regression could benefit adults. As far as childhood regression is concerned, the addition seems valuable and, as in most of these cases, serves to provide a gratifyingly firm biological foundation for a common psychoanalytic obser-vation; but regression in adults remains as much a mystery as ever.

If we leave the evolutionary viewpoint on one side for a moment and ask ourselves how Freud tried to explain adult regression, we see that he did so in terms of the concept of infantile sexuality. In his view, the fact of infantile sexuality left a sexual residue from childhood, as it were, one to which the unconscious could always return, especially if it were thwarted in the realization of more adult sexual aims. Here the concept of *transference* would come to the fore in suggesting that sexual experiences of the past could come to play a determining role in deciding the outcome of much later ones.

Basically, this is all transference is: the finding that there seems to be an unconscious, compulsive awareness of precedent by means of which contemporary experiences, relationships and situations are related back to earlier, typically infantile ones. A consideration of the possible evolutionary basis of transference in general must wait until later, but for the time being let us merely note that it is an important factor in sexual behaviour and one which naturally allies itself with regression.

It is also closely tied to the concept of *fixation*: the finding that the pleasure principle causes individuals not to want to give up gratifications which they enjoyed in the past. In developmental terms, fixations become sticking points to which the libido adheres

---

[5] For a brief account of Freud's theory of sexual stages in the context of modern evolutionary ideas see Badcock, *Essential Freud*, chapters 4 and 5.

and from which the EGO can have great difficulty in removing them. Typically, the most intractable fixations are found in childhood, as Freud's theory of infantile sexuality would lead one to expect.

Yet even this is problematic. Freud's observations convinced him that the effects of infantile sexuality on adult sexual performance were often negative: fixation on the sexual gratifications of childhood was often found to be a major factor in explaining the sexual inhibitions of adult life. If most sexual perversions were seen as a continuation of infantile sexuality into adult life and if their effect was mainly to reduce genital, reproductive primacy in favour of polymorphous perverse tendencies from childhood, such transferences from childhood seem paradoxical from the evolutionary point of view. This is because it is hard to see how something which reduces genital potency can promote an organism's ultimate reproductive success. Finally, when we include the paradoxes raised by homosexuality or transsexual behaviour, fetishism, narcissism and other apparent sexual aberrations, the whole question of adult sexuality, let alone that of the infant, poses seemingly insoluble riddles for evolutionary theory.

Oral manipulation of the mother

A surprising new perspective on the problem might be drawn from one of Freud's controversial findings about children – his proposed 'oral' phase of infantile sexuality. Freud thought that he was justified in calling this early manifestation of pleasure in sucking 'sexual' because it appeared to be enjoyed for its own sake, independent of whether the child was hungry or not. When he took into account the fact that such intrinsic pleasure in sucking, licking or kissing was a major factor in adult sexual activity and was on occasions promoted to the role of the principal pleasure in so-called oral sexual 'perversions', the case for regarding this early phase as the first expression of the libidinal drive in human beings seemed persuasive.[6]

According to the later findings of John Bowlby:

Though *sucking* is usually thought of as a means simply of ingesting

[6] Freud, *Three Essays* VII, 179–93.

food, it has a further function. All primate infants, human and sub-human alike, spend a great deal of time grasping and/or sucking a nipple or nipple-like object. In human babies sucking of a thumb or comforter is extremely common. In monkey babies brought up without a mother it is universal. When they are brought up with a mother, however, it is the mother's nipple that young monkeys suck or grasp ... to suppose that nutrition is in some way of primary significance ... would be a mistake. In fact, far more time is spent in non-nutritional sucking than in nutritional.[7]

Bowlby goes on to make much of what he calls 'attachment behaviour' in connection with sucking, but if sucking for its own sake is all about attachment, surely we would expect *oral* attachment to be less important to primates than to other mammals because of the fact that they have grasping hands and – excepting human beings – feet with which to attach themselves to the mother. Quadrupedal mammals by contrast, lacking such convenient attach-ing limbs, might be expected to make more of the oral point of contact. Yet, if anything, the reverse seems to be true and primates seem to suck even more compulsively than other mammals.

Furthermore, while his theory might explain why the offspring attempts to attach itself physically to the mother using the nipple as a convenient point of contact, it does not explain why the infant should *suck*, rather than merely *hold* it. And finally, if attaching itself to the mother were the point of the need to suck, why suck compulsively when separated? Surely, this is when sucking would *not* be necessary, rather in the way in which keeping one's eyes open is not necessary when asleep.

Although these are obvious criticisms of Bowlby's theory, it seems to have been much more popular than Freud's, perhaps thanks to the latter's unfashionable insistence on the libidinal element in oral behaviour and its connection with adult 'perversions'. Yet, from the perspective of the theory of parental investment, what Freud saw as polymorphous perverse infantile sexuality associated with the oral region or what Bowlby saw as attachment behaviour makes perfectly good sense as *an adaptation designed to secure maximum supplies of milk and other forms of investment in any particular infant at the expense of potential competitors.*

This is because evolution has already made the mother take one

[7] J.Bowlby, *Attachment*, p. 249.

or two enormous strides to meet the young offspring in its demand for the exclusive enjoyment of her investment by way of the fact that in human beings, as in many other mammals, frequent suckling serves to inhibit ovulation in females, and thereby conception of further, competing offspring. As long ago as Aristotle it was known that 'while women are suckling children menstruation does not occur according to nature, nor do they conceive; if they do conceive, the milk dries up.'[8]

In a recent paper, Blurton Jones and da Costa have suggested that night-time waking in toddlers may have evolved for this kind of reason.[9] Although they comment that 'a definite connection between crying at night and an increased tendency to suckle has yet to be demonstrated,' they also point out that there is some evidence that night-time suckling is especially critical to the contraceptive effect in question.[10] Again, there is also evidence that 'in the lactating woman, an infant's cry stimulates blood flow to the aureolar area and dripping of milk from the breast.'[11] However, they note in the closing paragraph of their paper that 'mechanical stimulation of the nipple, rather than depletion of milk, has been suggested as the stimulus that elicits endocrine response.'[12]

If 'the key to the short-term and long-term success of lactation as a contraceptive is therefore the frequency with which afferent neural inputs generated by the baby's stimulation of the mother's nipple reach the hypothalamus,' as a more recent review of the literature concludes, then it might pay any particular breast-fed offspring to continue to stimulate the mother's nipples as frequently as possible and for as long as possible, and not merely at night, but throughout the day. This is especially so since animal studies show that repeated stimulation 'not only helps increase milk production but may also provide additional contraceptive protection'.[13] Furthermore, studies of the chemical composition of human breast milk show that it is of

[8] Quoted in S. Thapa, R. V. Short and M. Potts, 'Breast Feeding, Birth Spacing and their Effects on Child Survival'.

[9] N. G. Blurton Jones and E. da Costa, 'A Suggested Adaptive Value of Toddler Night Waking: Delaying the Birth of the Next Sibling'. See also R. M. Nesse, 'Why Do Babies Spit Up.'

[10] Ibid., pp. 136 and 137.

[11] Quoted by B. Lozoff et al., 'The Mother–Newborn Relationship: Limits of Adaptability', p. 4.

[12] Blurton Jones and da Costa, 'Night Waking', p. 140.

[13] Thapa, Short and Potts, 'Breast Feeding', p. 679.

the kind characterizing continuous mammalian breast-feeders, rather than intermittent ones.[14]

If this is so, I see no reason arbitrarily to limit the effect to night-time suckling and to concentrate on waking as the critical behaviour, rather than on the compulsive oral behaviour described by Freud.[15] In my view Blurton Jones and da Costa are much more on the right lines when they recognize the possibility that 'nonnutritive suckling' in general 'may aid the infant in delaying the arrival of its next sibling.'[16] This is all the more compelling because, as a means of manipulating the mother's fertility, oral stimulation of her breasts by the baby seems to be extremely effective. Recent studies show that it is the duration of breast-feeding which explains 96 per cent of the variation in the persistence of infertility after birth. These studies also show that

a child's risk of dying in the first years of life depends on the preceding birth interval; mortality can be greatly reduced if it is born at least two years after its elder sibling ... The effect of preceding birth interval on child survival is found to be even greater than the effect of declines in parity or the mother's age at childbirth ... It has been estimated that a preceding birth interval of less than two years can raise the average chances of a child dying before age five by about 50% ... The best estimate of the true influence is that a second birth within 12 months of the index birth raises the risk of dying between the ages of one and five by at least 77%, if the mother is pregnant again by the index child's second birthday, this raises its risk of dying before the age of five by 55%.[17]

Again, Blurton Jones and da Costa quote Howell's data on the !Kung which suggests that 'lengthening the interbirth interval from 2 years to 4 years reduces mortality from over 70% to around 10%.'[18] If we calculate that the maximum four-year separation mentioned here requires an infertile period for the mother of about three years or so because the best part of a year will be taken up with conceiving and being pregnant to term again, it follows

[14] Lozoff et al., 'The Mother–Newborn Relationship', pp. 4–5.

[15] However, for a further discussion of a possible evolutionary background to night-time waking in infants see below pp. 92–4.

[16] Blurton Jones and da Costa, 'Night Waking', p. 140.

[17] Thapa, Short and Potts, 'Breast Feeding', pp. 681–2.

[18] Blurton Jones and da Costa, 'Night Waking', p. 137.

that infant survival is dramatically enhanced if the infant in question suckles compulsively for two to three years.

Although it is important to remember that evolution is driven solely by differential reproductive success and does not necessarily select for personal survival (recall the example of the guppy whose tail somewhat compromises his survival potential), the fact remains that, prior to an organism reaching reproductive age, survival is a necessary condition for any subsequent personal repro-ductive success. The consequence of this principle must be that if the figures quoted above for enhanced survival are in any way to be relied upon they strongly suggest that oral behaviour may have been powerfully selected because of the dramatic impact which postponement of sibling conception seems to have on existing infants. If we notice that compulsive stimulation of the mother's nipples for its own sake seems by far the most effective method of bringing about such spacing of births, then oral behaviour, far from being merely a means of securing some 'attachment' assumed to be of equal benefit to both parent and offspring, becomes one of the chief adaptive features in the human infant's competitive struggle for survival.

In the light of this consideration, it can hardly be coincidental that the crucial two- to three-year separation period for conceptions suggested by these findings is exactly the time-span allotted to the oral phase by psychoanalytic investigations. In this respect psycho-analysis once again reveals its characteristic sympathy for the child's, as opposed to the parent's, point of view. Well may parents resort to dummies, bottle-feeds or wet-nurses to outwit the child in its desire for exclusive enjoyment of its mother's breasts; and well may they enforce weaning and rationalize their self-interest by repression of their memories of their own oral behaviour and denial of the fact of infantile sexuality in general and its oral manifestations in particular. But the fact remains that young children do indeed have a special self-interest of their own in stimulating the breast; and it was Freud who first obtained a clear insight into breast-feeding from the infant's point of view.

Indeed, we can go further, for not only will stimulation of the mother's nipples above and beyond the child's need for food serve to postpone the appearance of competitors in the near future, it can on occasions become so compulsive an attachment that even if a new offspring should nevertheless appear, it may have to face considerable competition from an older child for access to the breast. For instance, among the primal hunter-gatherers of the central Australian desert, psychoanalytically inspired observations show that new-born babies sometimes die because older siblings will not give way to them at the breast, quite apart from the fact that their mothers will not enforce any kind of peremptory weaning of the existing child in favour of the newcomer. In this case, not merely before conception of potential rivals but after they have actually been born, the existing offspring has cause to want to monopolize the breast and finds what Freud called the oral attachment the means by which it is predisposed to do it.[19] Again, the observation that lingering oral attachments are often found strongly associated with the character trait of *envy* suggests that such factors as these can long survive childhood and can in certain persons influence behaviour decades later in adult life.[20]

If this interpretation of oral behaviour has anything to be said for it – and it does seem to be overwhelmingly vindicated by the facts – then it suggests that what may once have seemed to be the infantile equivalent of an adult perversion is in reality not in the least bit 'perverse' – at least in the sense normally attached to that term in relation to sexual behaviour. Far from being something which seems contrary to biological and evolutionary imperatives or something which is merely an expression of attachment behaviour, compulsive oral stimulation of the breast of the mother by the offspring for two to three years irrespective of immediate hunger seems a classical Darwinian adaptation.

Furthermore, interpretations of this kind can hardly be seen as serving the collective interests of the family, culture or species as mere attachment can. On the contrary, once the modern, individualistic approach of the theory of parental investment is taken into account and we allow ourselves to envisage the possibility of conflict, not merely between parents and offspring, but among offspring themselves, then the infantile oral period seems much less polymorphously 'perverse' than biologically natural, and much more a surprising product of evolutionary dynamics than one of apparent ethological commonplace.

---

[19] G. Róheim, 'Psychoanalysis of Primitive Cultural Types' p. 75.
[20] K. Abraham, 'The First Pre-genital Phase of the Libido'.

The Oedipus complex

But by far the worst instance of the apparent clash of Freudian findings and biological expectations is the case of the Oedipus complex. This, more than anything else, was found to inhibit, compromise and generally complicate adult, 'genital' sexuality with fixations, conflicts and repressions related to the infantile 'phallic' stage. Not only does this make the Oedipus complex look like a biological and evolutionary paradox, it also prompts disbelief because of its incestuous overtones. Incest seems to be contrary to the ultimate reproductive interests of the organism because of its deleterious genetic consequences, mainly thanks to its increasing the likelihood of an offspring receiving two copies of any recessive gene from its incestuously mated parents, thereby compromising its future.[21] Little wonder then that many concluded that Freud's ideas on the Oedipus complex had to be wrong, or, at least, seriously defective:

> Freudian theory claims that there are fundamental sexual conflicts in the human family. It is alleged that girls wish they had penises and boys are afraid fathers might castrate them because of their sexual interest in their mothers. On the surface, these kinds of conflicts make no obvious evolutionary sense. Selection against close inbreeding is usually powerful in nature, and it seems surprising that so dire a consequence as castration would need to be threatened in order to avoid mother–son incest. It seems more likely that Freud came upon sexual overtones in parent–offspring conflict and, lacking an evolutionary view of the relationship, misinterpreted the overtones for the real thing.[22]

In pursuing this line of thinking, I can suggest one respect in which the Oedipus complex could be seen to conform to the general expectations of the theory of parental investment. It may be that the situation with regard to Oedipal behaviour in childhood is comparable to the cases of regression or oral attachment, which also seem paradoxical at first sight because we might assume that offspring want to mature, rather than regress, and that they ought only to be interested in sucking when it would satisfy hunger. Of course, they do want to mature and they are concerned with food;

---

[21] See above p. 60.
[22] Trivers, *Social Evolution*, p. 146.

but Trivers's idea regarding regression is that the temporary benefits of less mature behaviour may well outweigh the cost to the offspring's eventual maturation. Similarly, we saw that in the case of oral libido, intrinsic pleasure in sucking may play an important role in ensuring the continuation of future supplies of food for a child always threatened with the possibility of new-born rivals for its mother's milk.

Similarly with Oedipal behaviour: perhaps it too is a form of parental manipulation, but one where the child manipulates the parents' psychology, rather than merely the mother's nipple. It may well be that its benefits to the individual child at the time it is manifested ultimately outweigh its eventual cost and that, as so often happens, such behaviour only looks paradoxical because we are not thinking about it in the right terms. Here the right terms are those of the actual costs and benefits to the individual child in its own right, rather than a presumed benefit to the species or family.

If we now ask how Oedipal behaviour could confer a benefit on the child by manipulating the psychology of the parent, the answer seems obvious. It seems possible that powerful feelings of love and affection directed by the child towards its parents might gain it very tangible benefits in terms of parental investment. It is by no means far-fetched to imagine that a child who shows deep love and devotion for its parents may receive some measure of love, care and affection in return. By showing how much it valued its parents a child might be more valued by them, and by trying to outdo its siblings in its love and devotion, an individual child might secure itself preferential regard by the parents in return.

This is all the more likely if we notice that Oedipal behaviour of this kind can be seen as an extension of the oral attachment which we have already discussed. This too, I suggested, was aimed at maximizing maternal investment. Nevertheless, studies show that orally induced infertility in the mother is not much increased beyond twenty-five months, meaning that any offspring will have to start resorting to other means of parental manipulation after approximately its second year.[23] This is presumably thanks to the fact that it is not in the mother's interests to allow the existing infant to close her options in this respect much beyond that time. Perhaps this is why oral manipulation gives way to the more general

[23] Thapa, Short and Potts, 'Breast Feeding', p. 680.

form of emotional manipulation represented by Oedipal behaviour and why the oral phase is found to fade at approximately this time, to be succeeded by the Oedipus complex between ages two and five. [24]

All this is by way of generalities, and I suspect that even the most abrasive critic of the idea of the Oedipus complex might concede that it had something in it. But there is more. Let us look at it from the point of view of the individual child or parent.

In early childhood, both sexes are equally dependent on the mother, who is the prime provider of childhood nutrition, care and protection in the vast majority of cases and especially so in primal societies. Freud rejected the term 'Electra complex' for the Oedipus complex of the little girl because psychoanalytic investigations showed that her relation to the mother was originally the same as the boy's and that her position was consequently far from a simple reversal or mirror-image of the latter's. However, as the child matures the father usually becomes more important and, by the end of a girl's childhood, he comes to exercise an important role in providing for his daughter's future – an observation as true in other societies as in primal ones.[25]

Since Oedipal behaviour is basically psychological in nature and exploits emotional and erotic appeal, it is not surprising that, as children of both sexes grow up, they should preferentially target the parent of the opposite sex. For the boy, there is no problem here. He targeted his mother initially and she remains the more likely to respond to his infantile amorous advances. The little girl, however, must switch to her father for exactly the same reasons which determine that the boy should retain his mother as the principal person to whom seductive behaviour is addressed. This is for the simple reason that the majority of parents are likely to be 'normal' in the sense that they are likely to respond best to emotional and erotic cues from persons of the opposite sex. The need for the little girl to acquire a new target in the cross-wires of her Oedipus complex after first acquiring the mother perhaps partly explains

[24] Freud, 'Two Encyclopaedia Articles: Psychoanalysis', XVIII, p. 245.
[25] For example, a recent study in the Caribbean showed that 'Young women with a father resident in the village are more likely to establish a stable mating relationship with a prosperous male than are young women without resident fathers.' (M. Flinn, 'Parent–offspring Interactions in a Caribbean village: Daughter Guarding' in L. Betzig, M. Borgerhoff Mulder and P. Turke (eds), *Human Reproductive Behavior*).

why the Oedipal behaviour of the little boy seems a simpler and more straightforward thing than that of the girl.[26]

Differential parental investment according to sex

But this is not the end of the matter where a boy is concerned. Another factor intrudes to emphasize Oedipal behaviour in boys and, as a consequence, to complicate further the psychology of the opposite sex. To see how this comes about we need to return to an issue touched on in the first essay. Earlier, in discussing the determinants of the sex ratio, I pointed out that, all other things being equal, it will not normally pay a parent to prefer investment in one, rather than in the other, sex. The time has now come to consider the situation when all other things are not equal.

Throughout that discussion I kept repeating the important qualification that all other things were equal, most especially in respect of my assumption that no particular parent had any way of knowing what the actual reproductive success of any particular offspring was likely to be. The reason I made this assumption was that it is self-evident that if parents were in fact in possession of information which could indicate the likely reproductive success of their offspring and if they could indeed determine their offsprings' sex preferentially, we would expect natural selection to favour those who made the 'right' decision about the allocation of their offspring's sex.

Here the 'right' decision is dictated by the basic principles set out in the earlier discussion. Most fundamentally, these are: first, that selection selects for individual reproductive success; and, secondly, that males nearly always show greater variance of reproductive success than do females, thanks to their numerous, 'cheap' and easily distributed sex cells. In other words, we can predict that parental investment in offspring will itself be subject to natural selection and that males, by general contrast to females, will tend either to have many more, or many fewer, offspring. As John Hartung recently put it, 'a man with ten wives can have more children than could a woman with ten husbands.'[27]

---

[26] For a fuller account see Badcock, *The Problem of Altruism*, pp. 47–56, and *Essential Freud*, chapter 5.

[27] J. Hartung, 'Deceiving Down', p. 173.

Going one step further, we might predict that

> males in better than average condition as adults may enjoy larger
> gains than females in better than average condition, while the reverse
> may be true when conditions are poor. Since condition in adulthood
> is affected by investment when young, parents may prefer to raise
> males when there is relatively more investment available per offspring
> and females when there is less.[28]

Such effects are likely to be particularly critical where offspring
sex is determined by environmental factors. In the case of birds,
incubation is usually carried out by the parents, and their body heat
(or, sometimes, shade) is an important investment in the young.
However, in the case of many lizards, turtles, crocodiles and alligators the mother merely buries the eggs in a nest and relies on the
ambient temperature to incubate them. Of course, such temperatures can vary widely as a consequence of weather and water
conditions, the position of the nest, its construction, and even the
relative placing of any particular egg within it. Unlike those many
animals whose sex is determined at conception by sex chromosomes,
sex-allocation in these species is now known to be determined by
ambient temperature. In the case of lizards, crocodiles, alligators
and the snapping turtle, above a certain critical level all eggs will
hatch as males, below it all will be females. In the case of other
turtles the reverse is true: higher temperature produces females,
lower ones males.

The point of view we are considering[29] would predict that environmental sex determination of this kind is hardly likely to have
evolved by accident, and that ambient temperature should be some
kind of reliable guide to future reproductive success. Since more
warmth means faster and greater growth during incubation and
larger body size at maturity, in each species we should expect to
find that the sex which gains most in reproductive success from
greater size should be the one determined by the higher incubation
temperature. In the case of crocodiles (and presumably that of
alligators too, although this has yet to be established) males have to
fight for females and so larger size seems an obvious advantage.

[28] Trivers, *Social Evolution.* p. 292.
[29] First put forward by R. Trivers and D. Willard, 'Natural Selection of Parental
Ability to Vary the Sex Ratio of Offspring'.

The same is true of the snapping turtle, but in the case of other turtles, larger size seems to favour females who consequently lay more eggs.[30]

Another example might be the potential effect of resource availability on parental investment. Studies show, for instance, that guppies fed a high-protein diet produce more males.[31] American opossums provided with extra food in the wild breed a higher proportion of males than control groups not so provided,[32] and in the United States 'people in high socioeconomic status tend to produce sons: they have about an 8% higher chance of producing a son than people at the bottom of the scale,' while 'people low on the socioeconomic scale have an almost 10% higher chance of producing daughters.'[33]

This is presumed to reflect the biased sex ratio at conception and the resultant circumstance that spontaneously aborted fetuses are correspondingly more likely to be male. Miscarriages are also presumably more likely the lower one goes in the socio-economic scale, thanks to declining standards of health care, increasing stress on mothers, and so on. But the basic principle can also be found where abortion is induced: for instance, in one study of a hospital in India all fetuses diagnosed male before birth were carried to term compared to only 5 per cent which were female.[34] Again, Dickemann showed in a famous study that sex ratios among Rajputs and other high-caste groups in India could become enormously distorted, thanks to preferential female infanticide.[35] These examples are presumed to reflect the fact that investment in males is risky, but justified if there is a good chance that the males in question might enjoy enhanced reproductive success.

Increased investment in females, by contrast, should be the preferred strategy in the contrary circumstances because females will almost always reproduce even when males might be at a disadvantage. This probably explains why 'female wood rats ... fed substandard diets appear progressively to kill male offspring by

[30] W. Gutzke and D. Crews, 'Embryonic Temperature Determines Adult Sexuality in a Reptile'.

[31] Trivers, *Social Evolution*, p. 286.

[32] S. N. Austad, 'The Adaptable Opossum'.

[33] Trivers, *Social Evolution*, pp. 297 and 300.

[34] A. Ramanamma and U. Bambawale, 'The Mania for Sons'.

[35] M. Dickemann, 'Female Infanticide and Reproductive Strategies in Stratified Human Societies'; 'The Ecology of Mating Systems in Hypergynous-dowry Societies'.

refusing them milk until finally they are nursing only daughters.'[36] The fact that 'women who have suffered or are suffering from schizophrenia tend to produce relatively more daughters (324 sons for every 417 daughters)'[37] presumably reflects this fact in the human population if we make the assumption that schizophrenia is a partly hereditary disease which reduces overall chances of superior reproductive success. Again, a study of German parish records of the eighteenth and nineteenth centuries suggests that whereas boys seemed to survive in higher numbers at the top of the social scale, significantly more girls survived at the bottom. Not only do data like these contradict neo-Marxist theories which hold that infant mortality reflects general economic conditions (and should therefore affect both sexes equally), they also strongly bear out one of the most provocative predictions of the theory of parental investment.[38]

In considering these examples, it is important to avoid one or two common errors. For example, it has been suggested that increased parental investment in males merely reflects their greater vulnerability and can simply be explained by use of the kind of reasoning set out earlier in relation to the determination of unitary sex ratios. This view proposes that because males wear less well than females they might be replaced more often, and, to the extent that they die off faster, investment in them will be enhanced.

In a sense, this is correct, but only if we recall what makes males wear less well than females. As we saw earlier, the factors which produce potentially greater variance in male reproductive success also account for increased male vulnerability. Fundamentally, this is the result of the characteristically greater male commitment to *mating success*, and the increased aggressiveness, risk-taking or possession of costly weapons or adornments which it entails. For example, if peacocks indeed wear less well than peahens, enhanced parental investment in peacocks would only be selected if it contributed to their parents' reproductive success, rather than merely to the individual peacock's longevity.

We must not forget that both males and females are nothing more than vehicles for the genes which they carry, not ends in

[36] Trivers, *Social Evolution*, p. 292.

[37] Ibid., p. 298.

[38] E. Voland, 'Human Sex-Ratio Manipulation: Historical Data from a German Parish', pp. 99–108.

themselves to be produced by parents for their own sake. Consequently, to say that increased vulnerability of males 'explains' biased parental investment in them gives only half an answer, leaving the question why they should be more vulnerable unresolved. A half-truth of this kind is certainly no 'perpetual alternative' to the explanation offered here.[39]

A second common misunderstanding probably arises out of a natural abhorrence of the idea of infanticide. Admittedly, 'murdering some of your offspring seems a rather drastic way to adjust the sex-ratio',[40] but only if we forget two important factors. The first is that parental investment is defined as anything which contributes to the reproductive success of an offspring *at the expense of the remainder of the parent's reproductive success*. Looked at from this point of view, selective infanticide might indeed promote the overall reproductive success of the parent, even if it cannot that of the particular offspring concerned. Secondly, we must notice that infanticide is only an option if parental investment continues significantly after birth, and that the greater the extent and the longer the period of post-natal investment, the greater will be the potential benefit of early termination if the outlook is unfavourable.

For instance, there are some species of bird which, despite the fact that they can usually only hope to raise one offspring per season, actually lay two eggs. Usually it is the first of these to hatch which gets most of the food, both because it can normally dominate its younger sibling and because its parents make no attempt to be equitable, but, if anything, favour the larger, older one. Almost always the younger dies, usually of malnutrition.

Perhaps to our eyes this seems callous and cruel, but, from the point of view of the parents, it does make sense. This is because the second egg is being used as a kind of insurance so that, if one fails to incubate or if the first hatchling dies, is malformed or sickly or does not thrive for any other reason, the second can be brought along as a substitute. Paradoxically, being able to raise only one offspring per season can make it important to start two, given that it is equally important to have one to raise.

It follows from reasoning like this that preferential parental investment in individual offspring after birth is likely to be especially

[39] S. Blaffer Hrdy, 'Sex-biased Parental Investment among Primates and Other Mammals: A Critical Evaluation of the Trivers–Willard Hypothesis', p. 134.
[40] P. Kitcher, 'The Animal Within: Biology and Social Science', p. 349

affected by the following factors:

1 the length of the period of post-natal investment;
2 the intensity and significance of that investment to the ultimate reproductive success of the parents and offspring;
3 the ability of parents to discriminate between offspring and to invest in them preferentially;
4 the ability of the offspring to solicit and/or manipulate the parents, their perception of the offspring and/or the resources available for investment;
5 the variance of potential reproductive success of any offspring related to sex, birth order, personal health, behaviour, or whatever;
6 the extent of the resources available to the parents to invest in the offspring.

If we look at each of these points in turn, it does not require laborious demonstration to show that almost all of them are especially pertinent to human beings. For instance, although human beings do not show the longest period of internal gestation of any mammal, they do exhibit by far the longest period of childhood and post-natal immaturity. Again, the intensity and significance of parental investment, and the ability of both the parents and the offspring to manipulate and monitor it, are vastly enhanced by human psychological and cognitive capacities. Admittedly, variance in potential reproductive success of individual offspring is not normally likely to exceed the range of other polygynous primates, but the last factor – the extent of the resources available for investment – means that what variance does exist is likely to be much more critical, or, at least, critical to much more investment than in most other species.

The 'sexy son' syndrome

The general principles set out above suggest one or two specific applications in the case of human beings. For instance, if a mother, shall we say, could estimate the likely eventual reproductive success of a child from evidence of behaviour during childhood, her own eventual reproductive success might be enhanced. For reasons which we have already noticed, such an effect is most likely to be found in the case of males, because individual females never enjoy such a wide variance of potential reproductive success as individual males do.

Considerations such as these prompt me to suggest that parents in general, but perhaps mothers in particular, might have evolved to become sensitive to evidence of potential reproductive success in individual males and that male offspring might correspondingly have evolved to produce it. In a polygynous species like human beings it is not difficult to predict what evidence might be relevant. Besides indications of general health and vigour it is possible that signs of precocious sexuality such as erections, amorous advances towards the mother and other females and aggressive behaviour directed towards the father and other males might be expected. The fact that this is exactly what we find strongly suggests that this effect, otherwise known as the male Oedipus complex, finds a further foundation in biological considerations – this time affecting not parental investment in general as I suggested earlier, but preferential parental investment in particular males.

In other words, I am arguing that Oedipal behaviour in little boys – we will return to the question of little girls in a moment – should be looked at from the point of view of its costs and benefits to the individual child in terms of parental investment. According to this way of looking at things, Oedipal behaviour would constitute a kind of 'rehearsal' or 'advertisement' of the boy's potential reproductive success, aimed at securing preferential parental investment because of the potentially much greater reproductive success which an individual male can have in the context of a polygamous[41] mating system.

From the point of view of the mother, we might begin to wonder where the notorious 'double standard' with regard to sexual behaviour really originates. This is because a woman shares half of her genes with her offspring (who inherit one complete set of chromosomes from her and one from the father), and also with her siblings (who received half of their genes from each of the same parents) and her parents (from each of whom she received half). On average, half of a woman's siblings and offspring will be male, and one of

[41] I say 'polygamous' and not 'polygynous' here in order to indicate that I include 'official' polygyny and extra-marital promiscuity since both are strongly implicated in a male's reproductive success.

her two parents was. If male philandering and gallivanting promote a man's reproductive success, they also implicitly promote the reproductive success of a woman who may be his mother, sister or daughter by approximately half that amount.[42]

The consequence of this is that a woman might be expected to have a somewhat different attitude to sexual adventurism in her father, brothers and sons than she does to the same behaviour in her mate. The latter is not normally closely related to her so that his philandering can only be at best irrelevant or at worst a real cost. But gallivanting sons, fathers and brothers can sow wild oats which contain approximately 50 per cent of the genes of a woman so related to them and thereby indirectly, but genuinely, promote her ultimate reproductive success. Perhaps this is why women often adopt astonishingly different attitudes to the sexual exploits of males to whom they are closely related by close ties of blood, as compared to those to whom they are related purely by marriage. If a double standard exists anywhere, it often seems to do so in women, as well as in men.[43]

The question which so perplexed Freud – 'What does a woman want?' – seems to admit of no simple answer if it is formulated as 'What does a woman want of males?' As we can now clearly see, a woman's biological, fitness-maximizing wants with regard to males are somewhat contradictory, depending on who those males may be, how closely related to them she is, and what the costs and benefits of their reproductive success will be to hers. Little wonder, then, that a mere male like Freud found the whole question so perplexing and its answer so elusive!

If we look at the case of sons in particular, we can see that here, more than anywhere else, the principle applies with a real vengeance. If a woman shares half her genes with her father, full brothers and sons she may be seen to have a vested interest in their reproductive success which, because they are all male, may –

[42] I say 'approximately' here because the measure being used is the crude 'simple weighted sum' measure of inclusive fitness, which is, strictly speaking, mathematically dubious if exact quantitative values are required. However, exact mathematics is not the point of this demonstration; what matters is the general qualitative point and here the 'approximately' can be interpreted to mean 'ignoring any effect of interference on a woman's reproductive success because of that of her near male relative'. See A. Grafen, 'How Not to Measure Inclusive Fitness'.

[43] I am indebted to my student, Tracey Fox, for making the full force of this argument clear to me.

especially in a polygynous mating system – show much greater variance than the range of reproductive success of a female, such as the woman herself, her sisters or her mother. But her son in particular is crucial here because he, unlike her father or her elder or contemporaneous brothers, is yet to begin his reproductive life, and is vitally dependent on her parental investment.

Admittedly, she may be able to help her father and brothers in various ways, especially if the latter are younger than she, but the ways in which she can help her own sons to reproductive success are likely to be much more significant merely because, as the mother, the woman in question is the prime provider of parental investment, at least in early life, and may exercise a decisive influence for much longer. This means that, as far as a mother is concerned, her investment in sons can, if they are reproductively successful, promote her reproductive success much more than any comparable investment she is likely to be able to make in daughters. 'Put more plainly, every highly successful male has a mother, and every mother with highly successful sons has an extraordinary number of grandchildren.'[44]

In species where the six factors listed above relating to the critical value of preferential post-natal parental investment are either absent or of relatively minor significance, such considerations as these are likely to be correspondingly less relevant. But in the human case where, as we have just seen, preferential post-natal parental investment is likely to be particularly critical – and perhaps especially so in the case of a mother with regard to her sons – these considerations may carry unusual significance.

In the case of oral libidinal attachments we saw that even though Freud described the oral phase as relating purely to the development of the child, the theory of parental investment opens up a new perspective on it and suggests that it is part of a larger picture which is actually based on the behaviour of the mother. If the mother did not inhibit her ovulation because of lactation in the way in which she does, the child would have no cause to exploit that adaptation by compulsive stimulation of the nipple with its mouth and tongue. Presumably mammalian mothers have this adaptation because it is in the mother's personal reproductive self-interest not to conceive again too soon after successfully giving birth to a new

[44] J. Hartung, 'Polygyny and Inheritance of Wealth' p. 5.

offspring. Having already made an enormous investment in the
neonate in terms of internal gestation, the mammalian mother's
own evolutionary interests coincide with that of the child in seeing
that she makes sufficient provision for it to begin to thrive. Conflict
between the wishes of the mother in weaning the infant and the
desire of the latter not to be weaned occur mainly over the question
of when weaning should occur, rather than the initial provision of
the milk.[45]

In other words, it is the termination of this particular phase of
parental investment which occasions conflict between offspring and
parent, not its existence as such, which is broadly in the interests of
both. Indeed, it is probably generally true that parent–offspring
conflict is most likely to occur most intensively at the margin of the
parental investment concerned, be it with regard to its timing,
quantity or quality.

Essentially my argument is the same for Oedipal behaviour:
Freud described the infant's side of the situation, but parental
investment theory allows us to evaluate the parental side as well.
This is an aspect of the situation which has not been visible in the
past, thanks in part perhaps to the interests of parents that it
should not be so. But again we see that early Oedipal behaviour in
both sexes and later Oedipal behaviour in males may correspond to
adaptations in the parents – perhaps particularly the mother in the
case of older Oedipal boys and fathers in the case of older Oedipal
girls – without which the infantile behaviour would make no
evolutionary sense. This is because unless parents responded to
Oedipal behaviour by providing greater marginal investment than
would be provided otherwise, it could hardly be accounted for by
the theory of parental investment.

Numerous observations seem to suggest that this is so. For
instance, among the Aranda aboriginals of central Australia Róheim
reports that mothers habitually sleep lying protectively on top of
sons until the latter are eight or nine years old, but does not
mention their daughters. If we assume that this behaviour is indeed
preferentially directed towards sons, as he seems to imply, and if we
look at it from the point of view of the mother, we can see

---

[45] Nevertheless, conflict could still occur at the margin represented by the
beginning of the investment if, as I shall argue later, post-natal depression in human
mothers is a case in point.

immediately that it makes sense and does seem to correspond to what might be taken as Oedipal behaviour in their sons.[46]

This is because the Aranda traditionally slept in the open, around camp fires. In the deserts of central Australia, as in deserts everywhere, night-time temperatures can be surprisingly low, and we have already seen that males are more vulnerable than females to all causes of death, disease and injury at all ages. From the point of view of her personal parental investment in her offspring, it might indeed pay a mother to discriminate in favour of sons where the provision of night-time body warmth was concerned because, not only would such sons be more vulnerable to cold, they might also be more worth preserving because of their greater variance of potential reproductive success. By contrast to the alligators mentioned earlier, it would seem that greater warmth in this case determines not so much that an offspring should be male, but rather that being male determines that warmth should be preferentially provided. Yet in both cases the underlying reality is presumed to be the same: the potential gains to parental reproductive success from discrimination related to the offspring's sex.

In our culture 'sleeping with' someone is a euphemism for sexual intercourse, but in the case of the Aranda it seems that sons might exploit amorous, Oedipal behaviour in order quite literally to get the benefit of sleeping with their mothers. This would not ultimately be because it was a symbolic fulfilment of a sexual wish – although superficially it certainly seems to be – but because a sexual wish directed at the mother made enhanced investment from her in the form of body-heat its actual fulfilment, to the direct benefit of the son and his potential reproductive success. Since his reproductive success is something in which his mother has a 50 per cent self-interest, thanks to the fact that he carries one half of her genes, some considerable convergence of mutual benefit can be discerned beneath this manifestly Oedipal sleeping arrangement.

Furthermore, since sexual intercourse between man and wife also occurred at night around camp fires (as well as in the bush in daytime – but that was often not between official sexual partners), it follows that a son who was favoured by his mother in this way was quite literally driving a wedge between his parents and perhaps physically hindering their opportunity for sexual relations. Indeed,

---

[46] G. Róheim, *The Riddle of the Sphinx*, p. 165; 'Primitive Cultural Types', p. 177.

Warren Shapiro reports that among the Miwuyt even in the daytime children will try to follow their parents into the bush, only to be restrained by others if it is obvious that the couple are going there for sexual intercourse. [47]

If this sexually intrusive behaviour could have the effect of postponing further conceptions and delaying the arrival of competing siblings then the Oedipal period, like the oral which precedes it, might have another important adaptive function, at least for the sons favoured with their mother's night-time embraces or toddlers inquisitive enough actually to follow their parents into the bush. Here antagonism towards the father would not be merely expression of future sexual competitiveness and masculine aggression, but an actual affront to his sexual opportunities dealt him by the person of his son.

Indeed, one cannot help wondering whether this is where night-time waking actually fits in, along with other features identified by psychoanalysis, such as the anal-sadistic phase which is supposed to follow the oral one from approximately age two. Quite apart from its undoubted significance for oral behaviour, night-time waking at a later age may perhaps relate to the infant's need to monitor its parents' sexual activities and perhaps to frustrate them whenever possible by crying, defecating and urinating, or even by just being seen to be awake.

It is certainly a common finding of psychoanalysis that Oedipal behaviour is intimately involved with real or imagined 'primal scene' experiences in which the young child is aroused by the parents' sexual activity and responds in some kind of active way. In one of Freud's most famous cases, that of the 'Wolf Man', the primal scene could be reliably fixed at one and a half years of age (in other words, towards the end of the time when oral behaviour alone is likely to control the mother's fertility) and seems to have resulted in the infant defecating.[48] Where children sleep in separate beds – or even in separate rooms – such interest in, and interference with, parental sexual activity looks relevant purely to themselves and has been treated so by psychoanalysts. Indeed, it is worth recalling that no less a person than the founder of psychoanalysis himself reports an incident from his childhood in which he wilfully urinated in the

---

[47] Personal communication, quoted by kind permission.
[48] Freud, 'From the History of an Infantile Neurosis', XVII, 36 & 80.

parental bedroom.[49] Nevertheless, Freud had the perspicacity to see that an evolutionary, innate tendency seemed to underlie such reactions – what he described as 'primal phantasies'.[50]

Yet in the conditions in which hunter–gatherers like the Aranda of central Australia live, with family groups bedded down together at night in intimate proximity around camp fires, infants are much more likely both to observe and to be able to respond to parental intercourse. Although there are good reasons for thinking that anal aspects of behaviour become exaggerated in other cultures and although it is undeniable that primal hunter–gatherers like the Australian aborigines are astonishingly uninhibited about excretion, it remains a fact that not even an aboriginal mother enjoys being urinated or defecated on by her children, perhaps least of all when she is in the throes of sexual intercourse with her husband!

To urinate or defecate on someone is universally understood as an expression of contempt and hatred and it is not difficult to see why Freud found that anal behaviour was 'sadistic' in its psychological connotations. Its evolutionary roots may lie precisely where those of the oral phase which precedes the anal-sadistic also lie: in the self-interest of the human infant in manipulating its mother's fertility and the inevitability of conflict between parent and offspring over this issue, at least for the first few years of childhood.

Significantly, Róheim reports that 'for the native Australian child only two things are really forbidden: to witness parental intercourse and to see the *churunga*.'[51] The latter represent male privilege and religious culture in the form of sacred engraved stones which may be seen only by adult male initiates; the former a prohibition which, presumably like all such prohibitions, represents parental interests in an instance of parent–offspring conflict. Again, he reports that parents only have intercourse at night if they think that the children are asleep.[52] If parents need to forbid children observation of intercourse in a culture otherwise notable for its tolerance of infantile sexuality of all kinds, the implication must be that children have a self-interest in observing (and possibly disrupting) it which challenges parental interests to the contrary.

[49] E. Jones, *The Life and Work of Sigmund Freud*, I, p. 16.
[50] Freud, *Introductory Lectures on Psychoanalysis*, XVI, 371.
[51] Róheim, 'Primitive Cultural Types', p. 178.
[52] Ibid., p. 177.

Although these observations may only seem to be true of the Aranda and comparable primal, nomadic hunter–gatherers, they probably have general significance for two very good reasons: First, we have already noted that primal hunter–gatherer prehistory accounts for the vast majority of the time our species has been evolving and that recent Australian aboriginal conditions may well be typical of it. Secondly, even among modern Western individuals, analysts habitually find that part and parcel of the Oedipus complex is a sensitivity to the possibility of parental intercourse and, very often, a latent desire on the part of children of both sexes – but sons especially – to disrupt it and put themselves between the persons of the parents. Although, analogously with the oral period, this has always been interpreted in exclusively sexual terms, it may well be that from an evolutionary point of view the theory of parental investment demonstrates wider, fitness-maximizing aspects which transcend the narrowly Oedipal aspects of such 'primal scene' fantasies with broader adaptive ones.

If this is the case, then it might generally be true that evolution would reward, not only regression and oral libidinal attachments in early childhood, but also preferential parental investment solicited by Oedipal behaviour in both sexes. Furthermore, this might be especially the case in human males who gave some intimation that they were likely to be reproductively more successful than others. If this were true it would provide a sound theoretical foundation for one of the most controversial observations of psychoanalysis: not merely the reality of infantile sexuality, but its fateful significance for adult life and its principal focus in the Oedipus complex of both sexes.

Penis-envy pays

Drawing such a parallel as this between the so-called 'sexy son hypothesis' in sociobiology[53] and the findings of psychoanalysis regarding the Oedipus complex immediately explains why Oedipal behaviour seems to be so much more obvious and unmistakable in males and why psychoanalysts find evidence of infantile penis-envy in females. The latter presumably reflects a female counter-strategy:

[53] P. J. Weatherhead and R. J. Robertson, 'Offspring Quality and the Polygyny Threshold: The "Sexy Son Hypothesis" '.

if males attempt to secure preferential investment in themselves by Oedipal behaviour centering on their potential adult sexual role as impregnators of numerous females, then little girls might promote their own eventual reproductive success if they were provided with something which made them respond positively, rather than merely accepting male privilege in this respect.

Presumably envy in general evolved as a part of the human behavioural repertoire because it serves to promote the self-interest of individuals and safeguard them against too ready compliance with the advantages enjoyed by others. Envy of what others have motivates one to secure the same advantage for oneself, and since evolution cannot always foresee what particular commodity might be in question, a general tendency to desire what others enjoy may be adaptive. Specifically, we have already seen that psychoanalytic investigations suggest that lingering oral attachments left over from early childhood are often linked to an especially marked development of envy. Presumably this is because oral behaviour has evolved to secure enhanced parental investment in the child and its continued expression in the adult suggests a frustrated desire for what others have had to their greater satisfaction.

Penis-envy appears to direct the little girl's attention to the defining characteristic of her male siblings: the possession of a phallus. The biologist Julian Huxley suggested the term 'psychological penis' for certain male sexual displays which seemed to have a direct effect on the sexual behaviour of females (such as triggering hormonal responses),[54] and the use of the term 'phallus' by psychoanalysts reflects a comparable observation. Although applied to the penis itself, the latter term does indeed see it as a *psychological* penis, one which is not necessarily reproductively potent, but psychologically immensely significant, and perhaps nowhere more so than in early childhood. Analysts found that it was the possession of this which the little girl envied. Although penis-envy was, in this sense, symbolic, and the penis envied evidently a psychological rather than reproductive one, it was found that in the concrete terms employed in the unconscious the penis had the consequence of turning its possessor into a male. Penis-envy therefore appeared to amount to envy of being male and becomes immediately comprehensible once we take the foregoing observations into account.

[54] J. Huxley, 'The Present Standing of the Theory of Sexual Selection'.

If preferential parental investment in males is in fact sometimes motivated by Oedipal behaviour on their part then penis-envy in females may well have evolved to counter it and to motivate little girls to want for themselves what otherwise might be preferentially invested in their male siblings. In effect it would be a simple behavioural predisposition to motivate immature human females to take account of what is in reality a much more complex and abstract phenomenon: preferential parental investment in males solicited by their Oedipal behaviour.

Although Freud could not have been expected to express the matter in the context of the theory of parental investment a generation before that theory was first formulated, one or two of his comments suggest that the link between penis-envy as such and the wider issue of what we would today call parental investment were apparent to him, at least as clinical observations. For instance, in his paper on female sexuality, he comments that at the end of the first, or 'pre-Oedipal' phase of attachment to the mother 'there emerges, as the girl's strongest motive for turning away from her, the reproach that her mother did not give her a proper penis – that is to say, brought her into the world as a female.' Nevertheless, and most significantly for the interpretation which I am now putting on this phenomenon, he then continues with the observation that a second reproach, 'rather a surprising one', often emerges in the same context. But it should come as no real surprise to us to learn that the reproach in question 'is that her mother did not give her enough milk, did not suckle her long enough ... did not feed her sufficiently.'[55] Reports from numerous parts of the world suggest that in many cultures such a reproach would be fully and literally justified: boys are indeed weaned later and breast-fed longer than girls.

One of my own students once provided comparable material. Following an account of the ideas given here, a lady in the class confided in me that she had no difficulties with the notorious concept of penis-envy. When I asked her why not she explained that, as a little girl, she had made a penis out of plasticine and placed it in her knickers! For her, the interpretations advanced here seemed obvious: looking back, she could see that her parents had really wanted a boy, had never given her the attention she thought

[55] Freud, 'Female Sexuality', XXI, 234.

she should have had, and that this was her way of trying to compensate.

Of course, as I pointed out earlier, the theory of parental investment predicts that preferential investment in males does not always pay. If parents face difficult circumstances, or have offspring which are less well placed to compete for mates, investment in females will be a better and 'safer' strategy because, as we have seen, female reproductive success varies less and is therefore more dependable. In a polygynous mating system all females will normally be mated, but the same does not apply to males.

This might suggest to some readers the necessity of some kind of symmetry, so that in circumstances where parents invested preferentially in females, males might develop a corresponding kind of envy. This would certainly pose problems for psychoanalysis, because no such phenomenon is observed or allowed for in Freudian theory. Fortunately, the problem is not a real one because the sexes are not symmetrical in this respect. To understand why this is so we need to recall the basic principle which says that, especially in a polygynous situation, males have greater variance in potential reproductive success than females. In other words, whereas nearly all females will mate, some males may not, and among those who do, some will leave many more offspring than others. What this means is that preferential parental investment in males is always somewhat speculative – a high-risk strategy which can go either spectacularly well or disastrously wrong, depending on the eventual performance of the male in question. Investment in females, by contrast, is a 'safe' strategy in the sense that it will almost always succeed in being moderately successful.

The consequence of this is that female offspring do not need to deploy especially provocative signals to solicit investment in themselves and therefore males who may be discriminated against do not have a similar provocation against which to react. Furthermore, it is important to remember that evolution is driven by reproductive success, not failure, and that, because of higher variance in male reproductive success in the past, subsequent generations are likely to be disproportionately descended from successful males, rather than unsuccessful ones. Special adaptations to favour the latter, such as a male equivalent of penis-envy, are consequently not to be expected.

Essentially the situation is like one in which high-risk investments tend to attract high interest rates, whereas lower-risk, more dependable ones earn more moderate returns. The earners of the high returns have no reason to envy those getting the lower ones, but the fact remains that the latter can count on much greater security in their investments than can the former and have their own reasons for preferring their admittedly lower rate of return.

Females are like the earners of the lower return; they have the compensation that even though they cannot expect huge profits, neither are they likely to have to absorb much in the way of losses. Their lower variance of reproductive success means that they are much more equitably represented among past reproducers and therefore much more likely to evolve and to pass on adaptations valuable to the majority of females in the majority of circumstances. Because preferential parental investment in males is a possibility at any time and an acute problem at certain times, penis-envy would certainly qualify as one of the adaptations which might be expected.

If resource availability and offspring quality is average, these considerations suggest that parents may do best to back their bets both ways and invest more or less equally in sons and daughters. In these conditions, sisters ought to have little reason to envy their brothers and penis-envy might be expected to remain a largely latent behaviour, present but inactive, like a computer program which is loaded but not being run. However, if conditions are especially good, parents might do better to invest preferentially in sons, and here their sisters should be ready to respond: penis-envy should not merely be *unconscious*, but should become *compulsive* – a behavioural program which is run, not merely stored. In the terminology suggested in the introduction, we might say that penis-envy ought to be seen as an inclusive-fitness-maximizing demand which is environmentally cued: just as alligator sex is determined by ambient incubation temperature, so penis-envy ought to be triggered by preferential parental investment in sons.

Furthermore, we should expect that female offspring who are competing with brothers who are in turn exploiting Oedipal, noisily masculine behaviour, might themselves do best to masculinize their response. Although good conditions should predispose parents to invest preferentially in males, masculine females might seem a reasonable second best if masculinity could enhance a female's ultimate reproductive success, for instance, in corresponding to a more assertive adult role for a woman which might ultimately promote that success. Little girls whose penis-envy predisposed them to masculinized behaviour might attract some of the investment otherwise earmarked for males. Indeed, in the absence of brothers such behaviour might attract all of it.

If resources are short or offspring quality poor, sisters should be motivated to compete even more actively because, from their point of view, let alone that of their parents, their brothers' potential reproductive success may represent a poor bet (assuming that such brothers exist). This suggests that penis-envy may be innate in all young human females, but particularly critical if they themselves rate their brothers as poor in quality or experience severe competition with them for resources which might indicate a lowered probability for male, as opposed to female, reproductive success. However, in the latter case masculinized behaviour on the part of girls would not pay, and so presumably penis-envy would manifest itself slightly differently: perhaps with girls being competitive towards brothers, but as females emphasizing their femininity rather than any masculine side of their behaviour.

In any event, since exact circumstances affecting the relative potential reproductive success of the sexes can seldom be reliably predicted in advance and independent of environmental conditions, we should expect penis-envy to be part of the behavioural repertoire of all females; at least in the sense that it remains a *latent* factor within the ID, ready to be stimulated into action should circumstances be appropriate. Psychoanalysis, with its unique method of research into latent factors, was bound to uncover it, and it is hardly surprising that it did so sooner, rather than later. But it seems that only now, with the coming of an evolutionary dimension to psychoanalytic theory, can we begin to comprehend its true nature and actual causes. Furthermore, the predicted link between the expression of penis-envy, environmental conditions and the pattern of parental investment provides a clear test for the theory and one which, although perhaps difficult to carry out, ought in principle to be possible.

However, considerations relating to resource availability and offspring quality do not tell the whole story because they fail to discriminate between the parents. Consequently, they fail to take account of the possibility that offspring could exploit differences in parental response to them. In so far as preferential parental investment in any particular male offspring is motivated by that male's

Oedipal behaviour directed towards the mother, a daughter might do best to try a similar tactic with regard to her father, so as not to let her brother have it all his own way.

In any case, an Oedipal male must express antagonism towards the father as evidence of his potential prowess in competing for females later in life, and this inevitably means that he cannot exploit quite the same relationship with his father as he can with his mother. This leaves such a relation with the father an open option for the daughter, at least in being her mother's competitor for her father's affections. In this way penis-envy and female Oedipal behaviour seem to be deeply linked and both seem to correspond to the same reality: the conflict introduced into female development by the possibility of preferential parental investment in males.

In mobilizing behaviour designed to respond to this situation, a little girl's ID communicates with her EGO via a symbolic code in which the physical reality of the penis stands for masculinity and in which envious feelings about it stand for envy at what masculinity can evidently mean in terms of preferential investment in males on the part of parents. In effect the EGO seems to receive a demand aimed at maximizing the little girl's inclusive fitness by making her sensitive both to the differences between the sexes and to masculine privilege.

Part of the resistance which Freud encountered regarding his discovery of infantile sexuality almost certainly arose because, from the adult, parental viewpoint, sex was 'unnecessary' and 'irrelevant' in childhood. Influential educational, psychological, sociological and political theories still take the same view. According to social-determinist dogmas (those which see culture or society as determining human behaviour), sex in childhood is purely 'gender' and determined by trivialities like what clothes a child is dressed in, what kinds of toys it plays with, what influences it absorbs from educational and other media, and so on.

But all such denials of the reality of infantile sexuality implicitly favour the prejudicial view of the parents and the culture, rather than the individual interests of the child. The ID, on the other hand, evolved to safeguard the interests of the individual's genes, and it takes a very different view of the matter. Having evolved in primal hunter-gatherer societies where preferential parental invest-ment in males is not just a possibility but a fact, it equipped children to exploit it, or to counter it, as the case may be (and as environmental and other conditions might dictate). The ID of little boys evolved to exploit the fact with Oedipal, precociously polygynous behaviour; that of little girls to counter it as best it could with envy of the preference shown her brothers and motivation to secure for herself by means of analogous Oedipal behaviour what would otherwise be theirs. If little Oedipus could drive a wedge between his parents and make a bid for his mother's preference, then his sister could do the same with respect to her father. After all, why should boys have it all their own way?

Contrary to the parental ideology represented by those who reject the fact of infantile sexuality and stress only the cultural, social or parental influence, evolution equipped human infants both to notice the reality of sex differences, and to react appropriately. As Robert Trivers commented in his original paper on parent–offspring conflict, 'in many species sex is irreversibly determined early in ontogeny and the offspring is expected at the very beginning to be able to discern its own sex and hence the predicted pattern of investment it will receive.'[56]

Looked at from the child's point of view, infantile sexuality is not 'unnecessary' or 'irrelevant' because not immediately linked to reproduction as it is in adult life. On the contrary, from the widely despised and neglected view of the child, sex is critical in infancy and infantile sexuality a strategic adaptation.

But psychoanalytic investigations have shown that penis-envy is not limited to childhood. On the contrary, it also appears to be a factor in adult life and may take two particularly significant forms. Either it may motivate a woman to compete with men in being masculine herself, or it may tend to make her see her male offspring as the fulfilment of her frustrated wish to possess the penis. As an aspect of masculinity on the part of a woman penis-envy appears to relate to the issue of homosexuality, consideration of which we must postpone for a while, but to which we shall certainly return. The alternative reaction, that which sees the son as a compensation, suggests a means whereby, if penis-envy in childhood could motivate competition for resources, its transference to adult life could also promote the very behaviour which it seeks to counter in a female's own infancy. In other words, it could be both the product and the

---

[56] Trivers, 'Parent–Offspring Conflict', p. 257.

cause of preferential maternal investment in males whose mother sees them unconsciously as a vicarious means of gratifying the desire for what she herself lacked in childhood. Here, once again, the penis plays a condensed, symbolic role, standing for everything that preferential parental investment in males suggests and closing a circle of causation which, if real, demonstrates the wonderful economy of means which evolution uses to bring about some of its most subtle and complex results.

The mother who idolizes her sons at the expense of her daughters is not a figment of literary or psychoanalytic imagination, but a real enough phenomenon to be noticeable in many families in everyday life. Indeed, in many cultures favouritism applied to males is positively institutionalized, with sons getting the best education, best food and conditions, and generally the first call on the resources which their parents have for investment in their offspring. For instance, it was entirely typical that the parents of Sigmund Freud should have provided him and not any of his sisters with the only private bedroom in their home (admittedly, a cramped one), and that when the first son of the family complained about the disturbance caused him by the girls' piano his parents should have had it summarily removed.

If feminists have sometimes been tempted to portray such preference as a dastardly male conspiracy to take over the world, they have nevertheless overlooked one notable fact. This is the realization that mothers often practise such favouritism with as much conviction as their husbands and that many women have traditionally seen male preference as both natural and desirable. On the issue of the 'natural', modern behavioural science might reluctantly have to agree – at least if that term is interpreted in an objective, rather than prescriptive sense (as it would be, for instance, if one were to say that 'death is natural'). As for the 'desirable', this is obviously a value-judgement, not a scientific statement which can be defended in any objective way.

In the past, Freudian insights into Oedipal behaviour and, most especially, penis-envy, have been contemptuously rejected and the latter seen as particularly offensive to women. But if we consider the possibility that preferential parental investment in males may be a fact, not merely underlying patterns of natural and induced abortion and infanticide among human beings, but contributing to their psychological adaptations too, then we ought perhaps to ask ourselves in whose interest it is to deny it.

Are the interests of women really best served by refusing even to consider the possibility that they may have been the real or potential victims of discriminatory investment? In whose interests is it really to further the myth that human parental investment is administered equitably and without prejudice? While I can readily see that it might serve the self-interest of some parents and some offspring, I can hardly believe that it serves that of anyone else. Once again, it seems that a synthesis of psychoanalytic and sociobiological perspectives reveals a seldom seen, minority viewpoint, but a valid one nevertheless.

Evolutionary causes of the castration complex

If the Oedipus complex was one of the most controversial findings of psychoanalysis, a further, subsidiary complex which is normally found at its core seems more provocative still: the so-called *castration complex*. Elsewhere I have given examples and provided a detailed description,[57] but here, in approaching the question from an evolutionary, adaptive viewpoint, we might do worse than to begin with the point just made: the perception of sexual differences in childhood.

For naked bipeds like our hominid forebears the observation of sex differences must have been simple. It seems that the unconscious ID adopts the same method as obstetrics does when it finds that it can reliably predict that a fetus revealed by an ultra-sonic scan is male if the penis in visible, but cannot predict with the same reliability if it is female. It seems that in ultra-sonic scans as in the unconscious, possession of the penis defines masculinity, absence of it implies femininity. In the symbolic code by means of which the ID communicates with the EGO 'male' means 'possessor of the penis' but 'female' is indicated by 'not-male', which in its turn is indicated by absence of the penis.

Furthermore, this is also the view of the child. Adults, with their prejudicial view of sex, may well think that sexual distinction turns on other, more important things than merely the possession of the penis, but the child knows better. After all, in childhood, behavioural or anatomical differences typical of adults of their respective sexes

[57] Badcock, *Essential Freud*, pp. 93–108.

have not as yet developed and so, just like the obstetrician in my example above, the child judges sex on one issue only: the visibility of the penis. This would have been unmistakable in our forebears, who, like the Australian aborigines of recent times, were almost certainly given to complete nudity, especially in childhood.

In part, this is what Freud meant by the term 'castration complex': the tendency to adopt a phallocentric view of sexual distinctions. Furthermore, it is not surprising that he has been heavily criticized for this by those who speak for the dominant, parental orthodoxy. But his finding was unmistakable and, if it was phallocentric in this respect, this was merely because the human infant and the human ID were equally phallocentric. Where identification of sexual differences is concerned, the evolution of parent–offspring conflict in our species has predisposed the child to take more account of the sight of the phallus than is accorded the often erroneous opinions of the parents.

Indeed, one might argue that in this respect the child and the unconscious are closer to objective, biological reality than are the parents and the conscious. This is because it is indeed a fact that human beings, like all other mammals, start out 'female' in the sense that they will develop as such unless they possess the gene (and it is indeed a single gene) for being male. In other words, on the most basic biological, molecular level, being male means having something which females do not have – an observation which makes the child's view of sex less irrational and subjective than it might otherwise seem.

Contrary to racist and socialist myths which attribute Freud's finding to racial or social 'conditioning', anthropologists who take the trouble to investigate this kind of thing come up with findings which can in no way be dismissed as the expression of typically 'Jewish' or 'bourgeois' prejudices. For instance, in the case of drawings by the Mehinaku indians of Amazonia, 'the penis is the major physical characteristic that differentiates men and women,' and 'in more than eighty percent of the sketches, male genitals are depicted as far larger than they are in real life.' In the view of the anthropologist who obtained the drawings, 'the male genitals were greatly exaggerated because my artists regard them as the principal element that distinguishes male from female.'[58]

---

[58] Gregor, *Anxious Pleasures*, p. 42.

Yet Freud's findings revealed that there was more to this than merely use of the male genital as a marker of masculinity in children, Amazonian indians and the unconscious. As the term 'castration complex' implies, it appeared that it was not just that the female of the species was seen as lacking the organ distinctive of masculinity, but that there was some presumption present that she had lost it. According to the 'surprisingly psychoanalytic' interpretations of the Mehinaku, 'female genitals are symbols of wounds ... The villagers' explanation of this ... is that a wound and female genitalia are similar in appearance. The edges of a clean wound come together, it is claimed, like the *labia majora* along the so-called genitals' path.'[59] In the case of the sex unfortunate enough to have suffered the fate of castration, what has already been said on the subject of penis-envy might go some way to explain this presumption. This is because a little girl needs, not merely to be able to distinguish sex reliably, but also to be motivated to feel that her own sexual identity might deprive her of something – evidently, parental investment – which she might have been able to command in certain circumstances had she been male. Here the idea of lacking something follows directly on the sex-defining concept and appears to have become 'condensed' so that two complex realities – being male and receiving preferential parental investment – seem to have been compounded into one compulsive realization: namely, that if one is female one has been deprived of what is otherwise only possessed by the male.

But if this theory explains the castration complex in the female of the species, it can hardly do so in the case of the male. Here it seems we must begin by taking other considerations into account, such as the following kind of objection which may well have occurred to many readers: 'Your theory is weak on two related points. First, there is no reason to think that apparent sexual behaviour in children is genuine at all, since, by definition, they are children and not adults. Furthermore, even if we took your idea seriously, what is to prevent a boy pretending to an eventual reproductive success which he will never in fact have?'

My answer to this objection is – paradoxically – that it fails to take the parent's point of view into account. Let us consider the following line of reasoning: if it would be in an offspring's self-interest to

59 Ibid., p. 153.

masquerade as something it was not in order to secure preferential investment from the parents, then it would also be in the parents' self-interest not to be fooled. This is because the parents might have other offspring who, if they were not so deceived, might receive the investment in question and promote the parents' ultimate reproductive success more than would otherwise be the case. In short, it will usually pay parents to detect such deceptions; and mothers who, for instance, invested preferentially in 'sexy sons' who only appeared to be such would lose in competition with other mothers who could detect the deception and invest more reliably, or who simply ignored it.

If it pays parents to detect deception in their offspring in the matter of the eventual reproductive success of males, male offspring must 'advertise' their potential success reliably. This means that the kind of Oedipal behaviour we are discussing is crucial to a male's later life. But how is he to know about that? It might pay him to know, for instance, whether he is really going to be successful in competition with other males. This is because, as I observed earlier, the costs of such competition are very high and the rewards usually somewhat uncertain. Yet the costs in terms of conflict with other males are only likely to be paid in adulthood, seldom if ever in infancy. Even if infantile sexuality is real, its antagonistic, aggressive aspect seems lacking, for it does not possess the vital reality element: costs which can only be incurred at a much later date.

Infantile castration anxiety might have evolved in little boys as an anticipation of these adult costs of sexual conflict without which infantile sexual 'rehearsal' of an adult sex role could hardly be real, reliable or credible to those for whose benefit it was contrived, let alone to the boys themselves. It seems to me possible that this is why Freud found that, in males particularly, the castration complex seems to imply, not merely that females lack the phallus, but that it has been forcibly severed from them – in other words, that a woman is a castrated man!

As it stands, this idea is absurd, just as penis-envy is absurd in itself. But, just as penis-envy makes sense if one considers it against its evolutionary background and the predictions of the theory of parental investment, so castration anxiety – for that is what we are discussing – makes considerable sense once we notice that, for a male, reproductive success and conflict with other males are normally inseparable realities of life and biology.

A myth told by the Mehinaku indians may reveal the reasons for this state of affairs. The hero insists on having sexual intercourse with a tabooed woman with the consequence that his penis grows to enormous size. So enormous does it become that at night it slithers around the village like a snake having sex with all the sleeping women. When the men see what is happening they beat the enormous penis until it is 'just a tiny one'.[60] Here the hazards which men associate with sex in this society, 'death, sickness, injury and failure',[61] are associated with the anger and aggressiveness of other men. The fact that such fears are often compounded with the object which arouses them – women in general, and the female genitals in particular – is only an expression of the fact that these anxieties are obsessive and operative in the unconscious, quite apart from their place in consciousness and their foundation in reality.

Psychoanalytic investigations of dreams demonstrate quite unmistakably that *condensation* of this kind is a principal feature of the unconscious, along with *displacement* – the means by which one, latent content can become attached to another, manifest one.[62] In the case of castration anxiety it seems that both mechanisms are evident: the fear of other men is condensed into a single entity with what occasions it, and thereby the anxiety about the one is transferred – or displaced – onto the other. When fears of this kind become compulsive and involuntary one suspects that evolution may have had a hand in shaping them and that some latent awareness of the dangers of sexual competition might have been selected and become an established part of the masculine ID.

If this is the case, then we might expect both that such compulsive fears would be present more or less from birth, and that their registration in the unconscious would associate them with early infantile acquisitions, consigned to more or less permanent repression by later development. This suggests that in childhood the compulsive awareness of the costs of sexual conflict with other males may well be expressed with a similar economy of meaning to that which we have just seen employed in the case of penis-envy.

[60] Gregor, *Anxious Pleasures*, pp. 132–5.
[61] Ibid., p. 151.
[62] Freud, *The Interpretation of Dreams*, V and VI; for a short account see Badcock, *Essential Freud*, chapter 3.

It may well be that the concept 'female' is represented by 'not-male' and 'not-male' by 'no-penis'. But it may also be true that, for males especially, the term 'not-male' can have an adverbial, as well as adjectival, meaning. From this point of view it would mean, not merely 'not being male', but 'not behaving the way males do'. If 'behaving like a male' means being aggressive and competitive with other males, then 'not being male' might connote a kind of behavioural castration – not being able to compete. Once again, radical condensation of the kind Freud first discovered in dreams and hysterical symptoms may underlie this entirely normal complex and explain why, for males especially, the ideas of femininity and castration are so closely allied.

Fundamentally, the castration complex in the male would mean much the same as it does in the female because in both sexes the crucial factor is identification of the offspring's sex in relation to preferential investment. If the argument set out above relating to penis-envy is correct, the loss the little girl really feels is not that of the phallus as such, but what it represents: preferential parental investment which might go to males because they are males – in other words, because they, rather than the little girls, did possess the penis.

In the case of a little boy, what he fears is much the same: his failure to secure preferential parental investment in himself as a potentially successful possessor of the penis. Yet he, unlike the little girl, has to confront a further danger in the the form of infantile castration fears which reliably represent the very real costs of sexual competition which he will inevitably encounter as a man. Failure to succeed in adult life will be a kind of castration because, as far as evolution is concerned, the only success that matters is sexual and reproductive – for a male, the reproductive success which he has in competition with other males. Little wonder, then, that in childhood the simplified symbolism of the unconscious should speak of castration as the danger always implicit in male sexuality and little further wonder that here again the child's view of things appears to be correct.

There would be all the more reason for this if males had more than one adult sexual strategy open to them so that, instead of being a 'regular', competitive kind of male, a man might be able to be a less competitive, or even non-competitive one. In his original paper on the theory of parental investment, Trivers remarks that if males within a relatively monogamous species are, in fact, adapted to pursue a mixed [sexual] strategy, the optimal is likely to differ for different males. I know of no attempt to document this possibility in humans, but psychology might well benefit from attempting to view human sexual plasticity as an adaptation to permit the individual to choose the mixed strategy best suited to local conditions and his own attributes.[63]

At first sight the suggestion that males might pursue different sexual strategies seems to contradict my assumption that masculinity and conflict with other males over access to females is inevitable. In reality, alternatives do exist and, as we saw earlier, there is more than one way to be a male. Furthermore, as we shall see later, the other types of human males in question might indeed be seen as undergoing a form of castration – at least in the sense that they compete by means of seeming to be less than completely male, or even female in appearance. In other words, *psychological* or *behavioural* castration may be what the child fears, so that castration anxiety is actually anxiety about future sexual role, rather than physical castration as such. It would be anxiety about reproductive behaviour in the future, first felt as anxiety about phallic sexuality in childhood.

In this respect it has a very clear parallel with something which we take as completely normal in childhood, at least in our culture. What I have in mind is *education*. Here, too, there is a great deal of parental investment at stake; here, too, the pay-off comes in adult life; here, too, a number of possible outcomes are open to children; here, too, children are set various kinds of anxiety-provoking tests which they must pass. The reason is obvious: the students must be tested because they too must know how well they are progressing; they must receive feed-back, encouragement and criticism from their teachers because so much is at stake and because so many possible avenues are open, not all of which any one student could possibly follow. Like infantile sexuality, education at school is a prefigurement of, and trial period for, the realities of adult life. In both, anxiety is inevitable simply because success can carry huge rewards and failure considerable costs.

Looked at from this point of view, childhood, and the Oedipal period in particular, might constitute a kind of trial period for a

[63] Trivers, 'Parental Investment and Sexual Selection', p. 146.

male's sexuality; one in which he could 'try himself out' and see what response he got. Here his mother's and other females' responses might be especially crucial, because they would show him whether he was regarded as a 'real', 'regular' male or not.

But his response to potential male competitors would be no less crucial, and here the whole weird, compulsive nature of the castration complex may find its explanation. Being regarded as a potentially successful male might involve, not merely impressing mother with his amorous proclivities, but giving evidence of being able to overcome castration anxiety aroused by father as well. If he thought that he was so regarded, then, for the reasons set out earlier, he might conclude that such an estimation had a high reliability rating and that he could take it that he had passed a test of his masculinity. Furthermore, the prefigurement of the costs, as well of the benefits, of sexual competition with other males in childhood might serve to prepare and forewarn the little boy of the very real risks which he will be taking when, as a fully grown man, he attempts to realize his polygynous ambition.

As far as a little girl is concerned, it seems that the greater 'safety' of female sexual strategies reflected in the fact that most females get mated in most reproductive systems means that the Oedipal period in particular should be much less crucial and Oedipal resolutions much less critical. This observation is broadly in agreement with psychoanalytic findings and may go some way to explain why analytic ideas relating to the male Oedipus complex appear to be much less easily applicable to female development than the formulators of the crude 'Electra complex' idea may have wished.

Paternity, identification and latency

If we return to the case of males for a moment, what I have just described is known as the *positive resolution* of the Oedipus complex – one which establishes a little boy's sexual identity as mainly masculine. But, clearly, the so-called *negative resolution* is also a possibility, and it is this which is thought to underlie many cases of apparent homosexuality in men: one in which the little boy's sexual identity is not confirmed as masculine (and which corresponds to the alternative sexual strategies mentioned earlier).

The problem is this: while preferential parental investment in 'sexy sons' makes evolutionary sense, and penis-envy and perhaps even pseudo-masculinity in little girls do also, pseudo-femininity in little boys seems not to do so. There seems to be no corresponding evolutionary pay-off for feminine behaviour by little boys in infancy and therefore no apparent explanation for certain kinds of adult homosexuality in men. Furthermore, the whole problem is made even more perplexing by the findings of psychoanalysis which show that adult homosexuality is always prefigured in infancy, albeit in very complex and far from well understood ways.

A by no means invariable, but significantly frequent, finding of psychoanalysis is that homosexuals of both sexes – but males especially – often report childhoods which featured weak or absent fathers.[64] The so-called 'positive' resolution of the Oedipus complex in men appears to be based on identification with an adequate father-figure, and the corresponding 'negative' or homosexual one on its absence.

One reason why, *identification* as such may be crucial in relation to fathers lies in the perennial uncertainty of paternity. Every breast-fed baby has plenty of opportunity to get to know the mother from the moment of birth, particularly by taste and smell. Kin recognition by chemical means is practically universal among animals; among mammals the habit of suckling the young gives offspring ample opportunity to sample the mother's characteristic taste and smell. In the human case, experiments have shown that breast-fed infants can distinguish their own mothers on the basis of smell alone after an exposure of about a week (but not before).[65] Yet such considerations can hardly apply to fathers. The small and numerous sex cells universally characteristic of males means that, whether it takes place by internal or external means, fertilization by any one particular male is usually only probable, rather than certain in the sense in which maternity is certain. From this it follows that paternity is not identifiable in the same way – that is, by taste and smell. Yet it can be established, at least probabilistically, by other means. So-called *phenotypic matching* is the means by which animals often assess relatedness. It consists of using the self as a standard of comparison, with the presumption that those who most resemble oneself are most likely to be kin and that the degree of relatedness reflects the degree of resemblance.

[64] Freud, *Three Essays*, VII, 146n.
[65] J. A. MacFarlane, *The Psychology of Childbirth*.

We know from psychoanalytic investigations that identification is a particularly crucial phenomenon for both sexes in relation to the father (and perhaps for the father in relation to the children[66]), but is especially so for males. 'Indeed,' remarks Freud, 'it almost seems as though the presence of a strong father would ensure that the son made the correct decision in his choice of object, namely someone of the opposite sex.'[67] In the words of Robert Stoller, 'in boys ... the more mother and the less father, the more femininity.'[68]

In non-analytic psychology too, 'there is a surprisingly vast literature on the alleged effects of father absence',[69] most of it strongly corroborative of the psychoanalytic findings. According to a recent review of this literature,

> several studies have shown relationships between the father's role in the family and the son's masculinity. Thus fathers who are seen as heads of households have more masculine sons and the masculinity of sons is lower when the father plays a feminine role at home. The greatest father–son similarity (in a variety of areas) has been found in families in which fathers dominate their wives.[70]

Fifteen separate studies showed that boys raised without fathers were less masculine and six others suggested that they may alternatively exhibit 'compensatory' hypermasculinity and aggressiveness. As psychoanalytic theory would predict, 'when the age of father–child separation is considered, studies show that father absence has the greatest effect on the masculinity of boys separated from their fathers in early childhood.'[71]

As far as homosexuality is concerned, numerous studies come to the same conclusion as psychoanalysis:

> An inadequate father-child relationship often appears to be a major factor in the development of homosexuality in males. . .There is much evidence that male homosexuals do not usually develop strong attachments to their fathers ... homosexuals who take a passive, feminine

---

[66] R. H. Porter, 'Kin Recognition: Functions and Mediating Mechanisms', p. 196.
[67] Freud, *Leonardo da Vinci and a Memory of His Childhood*, XI, 99.
[68] Stoller, *Presentations of Gender*, p. 25.
[69] H. Biller, 'Father Absence, Divorce, and Personality Development', p. 490.
[70] M. Lamb, 'Fathers and Child Development: An Integrative Overview', p. 19.
[71] Ibid., p. 27.

role in sexual affairs have a particularly weak identification with their fathers and a strong one with their mothers.[72]

It may well be that, since a boy cannot recognize his true father in the same way that he can his true mother, he relies on identification. This is important because, as we shall see in a moment, precisely the same reasoning might apply to young females with regard to choice of best mates as applies to young males in the estimation of their sharing the right stuff with their fathers. Here 'the right stuff' must be genes for successful polygyny which a son might inherit from his father. Provided with a reliable identification with a successful male, a young boy might feel some confidence in his own potential polygynous abilities and develop into a regular male. However, lacking this, and perhaps experiencing other counter-indications in his infantile interactions with females, a boy might be best advised not to adopt such a high-risk policy but instead to opt for a subordinate sexual strategy analogous to homosexuality.

If we enquire what kind of counter-indications might be important, one or two other observations might be explained. For instance, on purely theoretical grounds we might predict that physical characteristics related to sexual dimorphism (or the lack of it) might be a significant indicator. This would suggest that both the father's observed degree of sexual dimorphism and the potential degree of sexual dimorphism in the boy might correlate with final sex role to some extent. I know of no data which assess fathers' physical appearance as a function of son's final sexual orientation, but there is indeed evidence that homosexual men are significantly lighter in weight and weaker in muscular strength than heterosexual men.[73]

Again, Robert Stoller comments that, apart from the weak or absent father and the close tie with the mother, it is typical that the 'transsexual' (someone who not only wants to wear the clothes of the opposite sex, but *to be* one of the opposite sex) 'will have been perceived in infancy by his mother as being beautiful and graceful'.[74] This suggests that while reduced masculinity in appearance may be a factor in predisposing a boy to homosexuality, actual femininity might predispose him to an even more extreme form of 'transsexuality' in which masculinity is abandoned altogether.

[72] Biller, 'The Father and Sex Role Development', pp. 335–6.
[73] R. Evans, 'Physical and Biochemical Characteristics of Homosexual Men'.
[74] Stoller, *Presentations of Gender*, p. 34.

On the behavioural front, another indicator might be the level of parental investment. It follows logically that, if Oedipal behaviour in a male is intended to solicit preferential parental investment, especially perhaps from the mother, then the results of this strategy ought to be some measure of how well the behaviour had succeeded. However, apparently successful results would have to be carefully distinguished from dominant maternal behaviour (perhaps with high investment nevertheless) because maternal dominance might indicate failure on the part of the father to provide an adequate masculine role-model for his son. A research finding which tends to bear this out suggests that 'what seemed to inhibit the boy's masculine development was not the father's participation in some traditionally feminine activities in the home *per se* (e.g., helping in the housework), but the father's general passivity in family inter-actions and decision-making.'[75]

Other survey data indicate that 'a close-binding mother-son relationship seems more common in homes where the father is relatively uninvolved and may, also with related factors, lessen the possibility of the boy's entering into meaningful heterosexual relationships.' Furthermore, it seems generally true that 'maternal dominance has been associated with a varied array of psycho-pathological problems, especially among males.'[76]

However, analytic investigations have shown that maternal domi-nance can become an independent factor, being much more than merely an inevitable corollary of passivity on the part of the father. Furthermore, these findings seem to relate to predictions found in basic biological theory. In his original paper on parent–offspring conflict Robert Trivers mentioned the possibility that conflict could occur over the duration of parental investment in a manner exactly opposite to that mentioned at the beginning of this essay. In other words, the parent might want to continue to invest for longer and in larger quantities than the offspring itself would have wished:

> For example, where the parent is selected to retain one or more offspring as permanent 'helpers in the nest', that is, permanent nonreproductives who help their parents raise additional offspring (or help those offspring to reproduce), the parent may be selected to give additional investment in order to tie the offspring to the parent. In

[75] Biller, 'The Father and Sex Role Development', p. 327.
[76] Ibid., pp. 335–6 and 333.

this situation, selection on the offspring may favour any urge towards independence which overcomes the offspring's impulse toward additional investment (with its hidden cost of additional dependency).

He also remarks that 'it remains to be explored to what extent the etiology of sexual preferences (such as homosexuality) which tend to interfere with reproduction can be explained in terms of the present argument.'[77]

In the light of these remarks, it is significant that Robert Stoller reports that for a highly feminized male transsexual, 'his mother, not his father is the model for his gender identification ... He wants *to be* like her rather than *to have* her.' The mothers of such boys provide a 'lush', over-protective home environment and seem strongly to identify their sons with themselves. Typically, 'nothing is to split mother and son apart ... *mother and son are merged too well and too long*.' He concludes that 'it is this passionate motherhood that produces femininity or – to flip the coin over – that arrests the development of masculinity' to the point that, by age three, four or five, a boy 'may even announce that he wishes to have his penis removed'. Stoller goes on portray the son as the mother's 'perfect penis', giving examples of the role of penis-envy in motivating maternal investment which I mentioned earlier.[78]

It seems that what analysts like Stoller describe is exactly the effect predicted by Trivers: the finding that parents may identify their offspring with their own selves to the extent that those offspring are manipulated into giving up their own reproductive aims altogether. This comes about because a boy in this case identifies not with his father in the least, but with his mother in the extreme.

Furthermore, we can use this case as a clear test of the theory advanced earlier according to which Oedipal behaviour, whether in sons or daughters, is to be seen as a means of manipulation of the parents, aimed at securing enhanced arental investment. If we are also proposing that parent–offspring conflict can work the other way around, with parents wanting to provide more investment than offspring wish, Oedipal behaviour ought not to be observed under these circumstances.

This is all the more telling an expectation because, as Robert Stoller points out, analysts 'are used to finding Oedipal conflict as

---

[77] Trivers, 'Parent–Offspring Conflict', pp. 261–2.
[78] Stoller, *Presentations of Gender*, pp. 33, 30, 32 (Stoller's emphasis), 28–42.

the source of pathology' to such an extent that some of their critics maintain that they will find it everywhere, no matter what the truth. So it comes as all the more significant an observation that 'there is no Oedipal conflict' in these cases. In part, this may be a consequence of the weak-or-absent-father syndrome, but the same kind of effect seems to be seen in the contrary case of very masculine females. According to Stoller, 'if an excessively close mother–child symbiosis and a distant and passive father produce extreme femininity in males, too little symbiosis with mother and too much father could produce very masculine females.'[79]

In both instances what we may be seeing is a human, psychological equivalent of an effect not uncommonly seen in species where parent–offspring conflict over the offspring's reproductive role is resolved by the parent suppressing the reproductive physiology of the offspring. In numerous species, including some mammals, more dominant breeding females suppress the sexual cycles of those of lower rank. Perhaps in the case of the transsexual we see the human equivalent: suppression of the offspring's reproductive role by means of psychological manipulation which results in an unusually strong and exclusive identification on the part of the infant with the parent of the opposite sex. This would represent the exact opposite of a positive Oedipal identification: one which, far from confirming a child in its adult reproductive role by identification with the same-sex parent, in fact consigned it to a completely non-reproductive role by way of identification with the parent of the opposite sex.

Elsewhere I have proposed the idea that identification has a much more general role in human psychology.[80] This is because comparison with the self can also reveal the degree of likely relatedness between individuals and their kin. Normally, altruistic sacrifice will not pay an organism because we define such acts as *those which promote the reproductive success of the recipient at a cost to that of the performer.* Since evolution is driven by differential reproductive success, genes which promote the interests of others at a cost to their own reproductive success cannot – either by definition or in reality – be selected.

---

[79] Ibid., pp. 61, 33 and 57.
[80] Badcock, *The Problem of Altruism*, essay 1: 'Kin Altruism, Identification and Masochism'.

However, a gene for self-sacrifice in the interests of others could become selected if it conferred benefits on copies of itself in near relatives to an extent which exceeded the cost to itself in the individual making the sacrifice. For instance, since my offspring each receive one half of my total genes, it might be in the interests of those genes for me to sacrifice my life if it secured continued existence for at least three children fathered by me. Such a sacrifice would eliminate 100 per cent of genes residing in my body, but safeguard 150 per cent of them in my offspring, since each had received one half of my total complement of genes. Assuming that comparable ratios affect the likelihood of an offspring of mine inheriting my genes for self-sacrifice, altruism of this kind could evolve, all other things being equal.[81]

My suggestion is that identification is the characteristic means by which many acts of altruism in human beings are motivated, because *under primal conditions the extent to which one individual might identify with another would probably reflect the degree to which they actually resembled each other thanks to common inheritance.* In other words, most of those who reminded me of myself in a primal hunting and gathering society probably would be my relatives, an effect which would reach an extreme in an identical twin (with whom I would share 100 per cent of my genes).

Using a similar argument, Keith Sharp has pointed out to me that it might pay an offspring to sacrifice some amount of parental investment in itself if it had other siblings who shared enough of its genes to confer a net benefit upon them. He argues that resolution of the Oedipus complex in both sexes suggests some degree of mimicry of parental behaviour which could be directed towards younger siblings. Since such younger siblings would presumably be in greater need of parental investment than the older ones exhibiting the parental mimicry, the net benefit to shared genes for such mimicry and the tolerance of investment in younger siblings which it implies might exceed the cost to the individuals concerned. Consequently, such identification with parental interests could evolve.

However, such altruistic identification with the parents would only be expected in older children, most definitely not in younger ones. On the contrary, we have already seen that in early childhood

[81] See my discussion of kin altruism in *The Problem of Altruism*, based on a mathematical model first proposed by W. D. Hamilton, 'The Genetical Evolution of Social Behaviour'

siblings of similar age are severe threats to one another's survival and are arch-competitors for parental investment. But as a child grows up (and assuming that there are no siblings sufficiently close in age to be continuing rivals), both quantitative and qualitative demands on parental investment might change to the point that the cost to such a child of tolerating very much younger siblings and perhaps even identifying with the parental attitudes towards them might be less than the benefit. For instance, a fully weaned child of seven or eight years of age could afford to be much more philosophical about the mother's provision of milk to younger siblings than might have been the case some years earlier.

In any event, identification with the father and his values, ideals and interests (what Freud called the *superego*) seems predictable as a son grows older. This is because paternal – as opposed to maternal – investment seems to become more important the older a boy is, especially in a primal hunting and gathering society where a man's standing as a hunter is all-important. Even in the case of girls, reduced reliance on maternal investment might mean greater identification with the mother, rather than reliance on her as an object.

Freud called the developmental stage intervening between the Oedipal phase of childhood and adolescence the *latency period* and believed that it was characterized by a relative quiescence of infantile sexuality and an advance in development of the ego. According to the line of reasoning being pursued here it could be seen as a product of both a reduction and a shift in reliance on parental investment from the mother (exploited by a form of object-love) to the father or mother (normally by means of identification with the parent of the same sex as the child).

However, Sharp goes on to make the further point that the same argument suggests that children who are not full siblings should be tolerated to only half the extent of those that are because such half-siblings share one quarter, rather than one half of their genes. From this it follows that parental mimicry related to them should be reduced by a similar amount and Oedipal identifications consequently weakened, while the period of Oedipal rivalry and conflict over parental – especially maternal – investment might be lengthened for precisely the same reasons. Since half-siblings who are potential competitors for investment on the part of the same mother must be the offspring of different fathers, it follows that paternal identification must become complicated by that very fact.

In conclusion, I am inclined to agree with Sharp that

> it may therefore be possible to regard the whole matter of the latency period as the time when the child's chief concern – once its own survival is taken care of – is the survival of its siblings, which is achieved through the post-Oedipal identifications, and the relative inactivity of all the repressed conflicts and sexual desires of the Oedipal period. The re-activation of sexual aims at the close of the latency period would correspond to the child's new chief biological concern: its own reproduction ... Like most adaptations, its significance would obviously have been greater in the past, but the modern manifestation of it seems to be quite evident, particularly in girls, most of whom seem to act out maternal roles – and especially those related to child care – with particular enthusiasm.[82]

Latency would, then, correspond to a situation in which parental interests in recruiting older offspring as temporary and part-time 'helpers in the nest' partly coincided with the interests of the offspring themselves, perhaps especially in girls. However, this kind of partial coincidence of self-interest needs to be carefully distinguished from the complete suppression of the offspring's independent reproductive interest as found in transsexualism of the kind discussed earlier.

Considerations such as these would immediately explain why Freud found that childhood in general, but the Oedipal period in particular, played such a fateful, determining role in the development of adult character, sexual identity and behaviour. If it were indeed a period of trial and testing which formed the fundamental attitudes of the EGO towards its own self, then the later influence of early experience on the structure of the personality would be entirely comprehensible from the point of view of evolution and a major synthesis of Freudian and Darwinian insights would have been achieved.

Oedipal fixations figure

If we now ask what exactly these later influences might be, we could start by wondering what effect psychoanalytic investigations suggest

[82] Personal communication.

ʃipus complex has on the life of adults. By far the most
n is a tendency for adult love-objects to resemble the infantile
he mother in the case of a man, the father in that of a
woman.

Here we encounter the strange phenomenon of transference
mentioned earlier: the compulsive tendency to remodel present
experiences and relationships on the basis of past ones. In the case
of adult sexual relationships the decisive unconscious precedents
seem to originate in the Oedipal period of childhood and so
effectively our problem is to find some explanation for this most
paradoxical circumstance: why should *adult* sexual life take its cue
from childhood?

Adhering to the methodological approach advocated at the beginning, we might consider what consequences such Oedipal transferences could have had in primal hunter–gatherer societies and
look for clues in the only cultures which give any reliable guide to
them – those of the Australian aborigines.

Here, as mentioned earlier, we find that all societies are polygynous, some extremely so. In the case of the Tiwi, for instance,
thorough-going polygyny meant that 'all females must get married,
regardless of age, condition or inclination' so that 'there was no
concept of an unmarried female in Tiwi ideology, no word for such a
condition in their language, and in fact, no female in the population
without at least one nominal husband.'[83] However, in accordance
with the polygynous principle which so outraged Dr Arbuthnot, 'the
total female population, but only a part of the male population, was
married.'[84] Those Tiwi males who did have wives sometimes had as
many as twenty-nine, and 'as late as 1930, men with lists of ten, eleven
and twelve wives were still plentiful.'[85] Although, as one of the
accounts from which I am quoting immediately goes on to point out,
such numbers seem unusually high, even for Australian aborigines,
they do indicate the possibilities at what was almost certainly the top
end of the range of aboriginal polygyny in Australia.

Fathers, as was common among the Australian aborigines, had
the right to bestow their daughters in marriage, at least within the
broad limits set by the conventions of the kinship system. Contrary

[83] C. W. M. Hart and A. R. Pilling, *The Tiwi of North Australia*, p. 14; J. C.
Goodale, *Tiwi Wives*, p. 227.
[84] Hart and Pilling, *The Tiwi*, p. 15.
[85] Ibid., p. 17.

to the assertions of theorists like Claude Lévi-Strauss who, as we saw earlier, thinks that aboriginal kinship systems promote 'equality' through 'sister-exchange', fathers bestowed daughters

> generally speaking, where some tangible return was to be expected. Put bluntly, in Tiwi culture daughters were an asset to their father, and he invested these assets in his own welfare. He therefore bestowed his newly born daughter on a friend or an ally, or on somebody he wanted as a friend or an ally. Such a person was apt to be a man near his own age or at least an adult man, and hence perhaps forty years or so older than the newly born baby bestowed upon him as a wife.[86]

In other words, Tiwi girls tended in the vast majority of cases to marry men who were at least of their father's age and sometimes older so that 'the overall situation is best expressed by saying that no Tiwi father, except in the most unusual cases, ever thought of bestowing an infant daughter upon any male below the age of twenty-five'.[87] This would make her husband about forty when she married him. According to another account which strongly corroborates this, statistical data show the 'average age difference between wives and their first husbands as being 18.9 years'.[88]

Tiwi refer to first husbands of a girl taking her 'like a daughter' and 'growing her up': ' "I first married to Black Joe when I small girl. He grew me up. He made me woman with his finger ... " Her husband assumes a "father's" role',[89] and the father the husband's, symbolized in a ritual where her father presents her husband with a ceremonial spear which he – the father – has first placed between his daughter's legs.[90] Furthermore, not only were husbands likely to be of the same age as fathers, they were, as we have seen, likely to be friends and allies of their wives' fathers. They were most emphatically not likely to be boys of a girl's own age.

Although the Tiwi probably represent something of an extreme, even among the Australian aborigines, many other accounts of other aboriginal groups paint the same picture. For instance, in the case of the Miwuyt (*alias* Murngin), Shapiro gives the following figures for number of wives per man over and under the age of

[86] Ibid., p. 15.
[87] Ibid., p. 16.
[88] Goodale, *Tiwi Wives*, p. 67.
[89] Ibid., pp. 47 and 45.
[90] Ibid., p. 50.

forty: over the age of forty, three men had seven wives each; five had five; three had four; thirteen had three; twenty had two; and forty-two men of that age were monogamists. By contrast, only one man under forty had as many as four wives; three had three wives; three had two, and there were fifty-two monogamists. Bachelors with no wives numbered forty under the age of forty (of whom only one was a widower) but only four above it (of whom three had lost their wives and only one had never been married):

> In short, 90 men forty years of age or over have a total of 179 wives – a mean of almost exactly two wives per man; whereas 106 men under forty years of age have a total of 85 wives. Which is to say that men forty years of age or over constitute 46 percent of the adult male population and possess 68 percent of the adult females; whereas men under forty constitute 54 percent of the adult population and possess 32 percent of the adult females.[91]

Such findings are by no means untypical of aboriginal societies in Australia. A statistical analysis of a larger sample of Groote Eylandt aborigines showed that the average age of a husband was forty-two and that of a wife twenty-four. Similar discrepancies in age between polygynists and their wives are reported for aboriginal groups all over Australia and as far south as Tasmania.[92]

In putting these findings in their proper context, it is important to notice that it must generally be the case in any polygynous system that husbands tend to be older – sometimes very much older – than their wives. Even when young women have a generally free choice of husband as in our own, nominally monogamous societies, husbands are still typically somewhat older than their wives. Indeed, irrespective of society, a 'near universal of human marriage is that husbands are older than wives'.[93]

If we recall my earlier point regarding the importance of polygyny in the vast majority of human societies throughout the greater part of our evolutionary history, it is clear that the disparity in relative ages between wives and their husbands exhibited by the Tiwi, Miwuyt, Groote Eylandt and many other Australian aboriginal societies is in no way unusual or 'unnatural'. On the contrary, if our

[91] Shapiro, *Miwuyt Marriage*, pp. 76–7. Quoted by kind permission.
[92] F. G. G. Rose, *Classification of Kin, Age Structure and Marriage amongst the Groote Eylandt Aborigines*, p. 475.
[93] Daly and Wilson, *Sex, Evolution and Behavior*, p. 302.

physical adaptations, as evidenced by sexual dimorphism, differential life expectancy, testis size and so on suggest that we are basically adapted for polygyny, those same features also suggest a basic adaptation to the age disparity between husbands and wives found throughout human societies.

This realization casts a very different light on Oedipal transferences, at least in the case of young women. In the Tiwi case it would seem that a girl had no choice; her parents usually made the decision about whom she is to marry. But supposing that she did have a free choice, what would she decide then?

If, as we must suppose, behaviour evolves to promote the ultimate reproductive success of the genes an individual possesses, we might imagine that, compelled or not, a young girl might herself prefer to be married to someone who reminded her of her father in age and social standing. This is because polygyny has the effect of increasing the disparity in the reproductive value of males and females. In polygynous situations like that of the Tiwi, every female is mated to someone for the whole of her reproductive life (and, in that society at least, even beyond it), but this is not true of all males. Furthermore, a minority of males have many more females, and therefore many more offspring, than others. The Tiwi man who had had twenty-nine wives (admittedly, not all at the same time) must have had a very large number of offspring! There is certainly no doubt that statistical data from other societies shows conclusively that polygynously married males have markedly more offspring than monogamously married ones.[94]

As far as the ultimate reproductive success of any young woman is concerned, it would obviously be better for her to have sons who were themselves successful polygynists because such sons would be more likely to leave numerous offspring. In Trivers's words, 'by directly observing the choice of other females, a female can learn which males are likely to be attractive to the daughters of other females.'[95] Consequently, the best place to find genes for successful polygyny might be in existing, proven and successful polygynists: men, who, like her father, had several wives and numerous offspring.

[94] For instance, polygynously married Mormon church leaders averaged between 15 and 25 children compared to between 6 and 7 for monogamously-married Mormons. (Faux, quoted in Daly and Wilson, *Sex, Evolution and Behavior*, second edition, p. 284.)

[95] Trivers, *Social Evolution*, p. 348.

An unconscious, compulsive desire to want a husband who reminded her of her own father might therefore be entirely adaptive and wholly in the interests of a young woman's ultimate reproductive success – the final arbiter of evolution.

From the point of view of a daughter, the most significant female whose choice she should observe ought to be her own mother and, assuming her father seems an adequate and successful choice, it follows from Trivers's reasoning, quoted above, that she might let her mother's choice of husband be some kind of guide for her. This may explain why research shows that a girl's, as well as a boy's, success in establishing a secure sexual identity is dependent on an adequate father, not to mention the fact that 'the effects may remain unobserved until adolescence.'[96] For instance, a study by Fisher found that 'paternal deprivation in early childhood is associated with infrequent orgasms among married women.' Again, 'inappropriate or inadequate fathering seems to be a major factor in the development of homosexuality in females as well as in males.'[97]

Although the reason for the latter observation cannot be quite the same as it is in the male case, it does suggest that paternal transferences are as important in their own way to the formation of a feminine sex role in adulthood as they are to a masculine one. Essentially, my suggestion is that whereas a boy identifies or otherwise directly with his father in constituting his potential adult sexual role, a girl identifies both with her mother and with her mother's choice of husband in constituting hers. Either way, it seems that fathers have much to answer for in the matter of their offsprings' final sexual identity.

In general terms, it seems that Oedipal behaviour, understood as a compulsive desire to model adult erotic ties on childhood ones, might not be as maladaptive as it may seem at first sight. On the contrary, in conditions of primal polygyny comparable to those reported for the Tiwi it could actually promote the reproductive success of those who carried the genetic determinants for such Oedipal transferences within them.

So much for women, but what of men? Surely no such Freudian–Darwinian synergy will be found in this case? Let us consider the

---

[96] Lamb, 'Father and Child Development', p. 28.
[97] Biller, 'The Father and Sex Role Development', pp. 346 and 347.

consequences of what we have just seen. In the case of the Tiwi, 'a ... husband was unavoidably and necessarily always much older than a bestowed wife. Therefore he usually died much earlier than she.'[98] Because no woman could remain unmarried, the widows were always recycled, very often to younger men who, while not important enough to established polygynists to warrant promises of their daughters, might nevertheless receive a 'second-hand' wife in the form of a widow. Since such widows, unlike younger women, could frequently exercise a considerable degree of choice about their new husband, a male responsive to the somewhat older woman might have some advantage in attracting such mates, assuming that younger ones were not available to him. At the very least, these observations show that, contrary to first appearances, male and female situations in this respect are not as different as might have been thought. On the contrary, the inevitable tendency for older men to marry younger women in any polygynous system means that older wives and younger, unmarried men must become potential sexual partners for one another.[99]

Nevertheless, it must be admitted that if this were the only respect in which a preference for maternal women affected the reproductive chances of younger men it would be a weak argument indeed, and in no way fully comparable to the parallel female case. However, there is another dimension to the situation which vastly strengthens younger males' biological motives for being interested in older women.

If we return to the question of female parental investment for a moment, we can see that it might pay a young woman to be 'coy' about sex in the sense that she should carefully weigh the costs to herself and her ultimate reproductive success of any mating offered by a male. This is not merely because of her vastly greater degree of potential parental investment should she become pregnant, but because she is at the beginning of a career of such investments, and early mistakes could be costly. For instance, matings with males who have not been carefully chosen may result in unfit offspring or progeny who might themselves be at a disadvantage when it came to their own reproductive careers. Again, as Robert Trivers points out (in the context of a discussion of parental investment in ring

[98] Hart and Pilling, *The Tiwi*, p. 18; Goodale, *Tiwi Wives*, p. 67.
[99] Rose, *Groote Eylandt*, pp. 92–104.

doves), 'we expect some ambivalence in males toward easily aroused females,'[100] and there is every reason to suppose that the principle applies to human beings too. This is because females who are insufficiently 'coy' may reduce a male's confidence that any off-spring conceived are his own and thereby compromise his wish to invest in them, to the obvious detriment of the female and her offspring.

Of course, the estimations regarding the suitability or otherwise of males which a young woman makes need not be conscious; although conscious thoughts related to the relative 'attractiveness' or otherwise of possible mates will figure in her deliberations. But evolution will probably have equipped her with more 'instinctive' feelings and with a sexual physiology geared to relatively slow arousal, a tendency to recalcitrance and general choosiness about the whole matter of sexual involvement.

For an older, but still nubile, woman, however, the situation is likely to be quite different. Physiological and behavioural studies show that older nubile women become aroused more rapidly, achieve orgasm more readily and with greater frequency, and consequently seem to experience what Freud called 'considerable increase in the production of libido.'[101] Perhaps the best evidence of this from the behavioural point of view relates to masturbation because, as Kinsey noted, 'the frequencies of masturbation depend primarily on the physiologic state and volition of the female' and therefore 'provide a significant measure of the level of her interest in sexual activity'.[102] He found (and more modern studies corroborate) that the percentage of women who masturbate to orgasm increases from about 30 per cent at age seventeen to 60 per cent at forty, with frequency peaking at the same age. In males, by contrast, both proportion and frequency drop progressively from about age eighteen. Observations like this suggest that for the older woman being 'coy' does not pay to anything like the same extent. Nor should it, because, as a woman approaches the end of her reproductive life, her reproductive value must decline. With less to lose she has more to gain from taking risks.

If this is true of her reproductive value to herself it is even more true of her value to her mate. If, as is likely, he is a polygynist, the

[100] Trivers, *Social Evolution*, p. 265.
[101] Freud, *Introductory Lectures on Psychoanalysis*, XVI, 403.
[102] A. Kinsey et al., *Sexual Behavior in the Human Female*, p. 146.

older woman will find that increasingly she has to compete for her husband's attentions with younger and more attractive wives.[103] Admittedly, the many psychological and social advantages which age and experience give an older woman equip her to carry on the fight for some considerable time. But the fact of menopause suggests that, eventually, the costs of sexual competition begin to outweigh the benefits to the individual woman to the point where she ceases to compete and switches her efforts towards investing in her existing offspring and kin.

However, between the point at which menopause intervenes to call a halt to her reproductive life and that at which she may begin to appeal to her official mate somewhat less than younger wives, a woman could find that alternative sexual partners were by no means impossible to find. Whereas young girls might be accounted for and, in any case, coy about risky, illicit and casual matings with unproven, young and as yet unmarried men, older but still nubile women could take a different view.

This would be especially so if, as seems likely, a senior wife could count on her husband's continued economic support more than on his sexual interest.[104] Secure in her position in her family and community, with grown-up children and other kinsmen to protect her and a socially, if not sexually, dominant position over other women in the household, an ageing woman might feel that she could look elsewhere than to her husband for impregnations.

More than this we may not be able to say as far as such a woman's sexual choices are concerned. Furthermore, I must point out emphatically that we are considering the situation only as it is found in what I take to be vestiges of primal hunting and gathering societies, with high degrees of polygyny and more or less immediate marriage of young girls to older men. In our own culture no such conditions exist – indeed, on the contrary, in our pseudo-monogamous system young women usually pass through a period of effective promiscuity in their youth which contrasts strongly with what may appear to be a more discriminating and 'coy' attitude to sexual encounters later. But here it is important to point out that modern conditions complicate sexual behaviour enormously and

---

[103] Goodale, *Tiwi Wives*, p. 227.

[104] In the case of the Miwuyt, for instance, Warren Shapiro reports that polygynists show a preference for sleeping with their younger wives (personal communication).

that in our societies considerations relating to economic factors may well figure more prominently as a woman ages and reduce the appeal of purely sexual factors.

Nevertheless, we must not forget that, as far as the evolution of human behaviour is concerned, modern conditions count for next to nothing, and that it is primal hunter–gatherer prehistory which has been the principal adaptive context in which human sexuality has evolved. Furthermore, if we return to a consideration of primal conditions we can see that this line of reasoning has important implications for males. It suggests that not only might young, unmarried men encounter available older women as 'recycled' widows, they may also find that many married women past their prime were potential sexual targets. Not only is this corroborated by female sexual psychology and physiology with their greater orgasmic potency later in life, it could also be evidenced in the corresponding sexual physiology of the young male.

Here, the relevant consideration must be the fact that young males become aroused and ejaculate much faster and somewhat more frequently than older ones. If we make the logical deduction that many, if not most, of the matings suggested by the argument immediately above would be illicit ones, carried on by older married women behind their husbands' backs, then it follows that natural selection might favour fast ejaculation in the circumstances. Always prey to the possibility of being caught *in flagrante* with the lady in question, it might pay a young man to become aroused and ejaculate rapidly since the likelihood of detection and disturbance must increase with the time taken. This observation is true of many mammals with polygamous mating systems and it is certainly true of feral sheep where 'the small ram ... has little time to copulate before the large ram arrives and butts him off the female' so that 'there is a steady selection for quick copulation.'[105]

In the past there has been a tendency to see a mismatch between male and female sexual physiology, with young males tending to a pattern of fast response found in older females, and with the opposite applying to younger females and older men. Almost certainly, this was the result of naive thinking about the sex ratio fitting each male to each female of a similar age with the whole thing being contrived for the benefit of the species as a whole. We

[105] Geist, *Mountain Sheep*, p. 222.

now see that, on the contrary, there appears to be a natural matchin of the respective responses of the sexes so that older women make natural partners for younger men in the context of a polygynous system where older men tend to be mated to younger women and where selection operates at the level of the individual, rather than that of the group.

Psychologically, this situation suggests Oedipal transferences just as naturally, with young men being predisposed to target older, 'maternal' females as potential, if usually illicit, partners, and younger women tending to target older men. Far from being 'unnatural' or 'paradoxical', Oedipal transferences of this kind might be seen as exactly what should be expected if evolution had a hand in forming them. Nor would the transference of infantile Oedipal wishes into adult life be at all surprising if their consequences were as I am suggesting here.

Of course, it is important to remember that, at least in the case of young men, what we are discussing is a distinctly second-best, subordinate or 'alternative' mating strategy. There can be no doubt that, given a free choice, most men would probably prefer more youthful wives. After all, this is what a young man ultimately aspires to, even in an extreme case of polygyny, as in the Tiwi. He hopes to become a successful polygynist himself and receive young women, rather than recycled widows.

In so far as this is the accepted, 'official' and 'regular' sexual career of a young man he has nothing to hide from his peers or elders. They will expect him to want youthful women and some of them may provide them for him in due course. What they most decidedly will not want is that their existing wives should become sexual targets for younger men. Since any such matings must be covert and dangerous, it might pay to hide such motivations and so evolution seems to have seen to it that Oedipal transferences in younger men are unconscious and compulsive, rather than conscious and voluntary.

To the extent that sexual interest in maternal women has to be concealed, perhaps even from the young men themselves, it might be seen as a psychological manifestation of cryptic sexuality: a subordinate sexual strategy which succeeds best if hidden. As a purely psychological adaptation, its hiding would itself have to be purely psychological, and hiding the contents of consciousness from consciousness is just another way of saying that such adult

Oedipal transferences would tend to become *unconscious*. To this extent consciousness of sexual motives would itself have become cryptic: a camouflaged consciousness not fully conscious of itself, a mind which appeared to be thinking one thing while actually thinking another.

Although I have used the Australian aborigines as my examples, many other societies could have been chosen and, clearly, the principle should apply to polygyny generally. This is because polygyny implies that some men – almost always older ones – will have a number of wives and that some other men – usually younger ones – will have to do without any as a result. As we saw earlier, some sociologists justify this inequitable state of affairs by claiming that what is good for polygynists is good for the society as whole.

Others have claimed that Oedipal triangles are purely culturally conditioned; and it is ironic to find that one of the most notable of these cultural-determinist sociologists, Bronislaw Malinowski, while claiming to refute Freudian theory, in fact found a culture which richly illustrates my point. This was the matrilineal Trobriand Islanders, where the mother's brother was supposed to be so prominent an individual that the Oedipal, father–son–mother triangle could not be discerned. Yet, as Melford Spiro pointed out in his re-evaluation of Malinowski's findings,

> almost all marriages in the Trobriands are monogamous, polygyny – a mark of wealth, power and prestige – being practiced only by chiefs. That the sons of chiefs are as adulterous as the sons of commoners is not surprising. What is surprising, however, is that they typically seek (and find) their paramours from among their fathers' wives (excepting, of course, their own mothers). This, surely, is a most extreme example of a son's fulfilment of his Oedipal wishes. To be sure, since his lover is only his step-mother, their affair does not, in the literal sense, constitute mother–son incest. Since, however, she is his father's wife, and therefore the structural equivalent of his mother, she is certainly, psychologically viewed, her symbolic representation. The woman's husband, however, far from being a symbolic representation of the father, is the son's actual father. Hence, the Oedipal triumph over the father that is inherent in such an affair is actual rather than symbolic.[106]

---

[106] M. E. Spiro, *Oedipus in the Trobriands*, pp. 103–4. I am indebted to D. Rancour-Laferriere's *Signs of the Flesh: An Essay on the Evolution of Hominid Sexuality*, p. 139 for reminding me of Spiro's finding.

It also suggests that the general point about polygyny which I am making can – and apparently sometimes does – have more specifically Oedipal applications and that the older women targeted by younger men for their sexual adventures are not merely of their own mother's generation, but are among their own father's wives.

In so far as Oedipal transferences can be seen as regressions to infantile sexual objects, we can begin to give some sort of answer to the riddle of regression in adult life. Far from contradicting basic biological insights, such regressions to infantile, Oedipal sexual targets seem to be in accordance with the fundamental fitness-maximizing demands of the organism, at least as long as regression is understood as conditioning adult sexual transferences and as long as we direct our attention, not to parental investment as such (the operative factor in infantile regression), but to the sexual lives of adults.

Although by no means a complete explanation of every kind of regression found in adults, these insights suggest a promising line of inquiry: namely, consideration of the role of regression in constituting cryptic adult sexual behaviour through transference effects from the individual's childhood. It is to a fuller examination of these effects that we must now finally turn.

Homosexuals as cryptic heterosexuals

Oedipal behaviour may evolve in males in infancy as a human equivalent of the so-called 'sexy son' effect, and, in infancy, competitiveness directed towards other males such as the father might not seem worthy of much serious concern. After all, a child is a child and can hardly constitute much of a real threat to a grown man. Threats in fantasy and parricidal wishes in infancy may not amount to much that an actual father need worry about; but, as a boy matures, the picture changes. In adolescence in particular, antagonistic behaviour towards other males may take on a more serious side and cease to be something which adult males can safely ignore.

In other words, the costs of aggressive Oedipal behaviour to the individual male will probably increase with his age, so that what he can get away with as a child becomes increasingly costly to him as a man. This suggests that some kind of repression, or forgetting and suppression of Oedipal behaviour, could pay as a boy grows into a man – an observation which provides another cogent reason why the repression of the infantile Oedipus complex should be a universal finding.

But it also suggests another possibility. This is the possibility that some males might avoid or reduce the costs of conflict with other males by behaving or appearing to be less male, perhaps even to the point of appearing to be female. Such a possibility corresponds to the 'alternative' sexual strategies mentioned earlier as possibly prefigured in infantile sexuality and associated with castration anxiety because they represent 'castrated', less masculine behaviour. It certainly agrees well with the ethological examples mentioned earlier and with the observation that 'alternative patterns are often "submissive", and can include the display of female characteristics which serve to reduce male aggression.'[107]

This realization immediately begins to make sense of the so-called 'negative' resolution of the Oedipus complex understood as an infantile predisposition to homosexuality, femininity and generally submissive and non-masculine behaviour. It suggests that if a little boy is not confirmed in a regular masculine character identity by the experiences of the Oedipal period, identification with his father or whatever, he might instead opt for a second, subordinate sexual strategy, rather as irregular males of the other species mentioned earlier do. The difference is that whereas evolution has differentiated the three types of the male bluegill sunfish as physical adaptations, it appears to have differentiated human males into a number of types as purely *psychological* adaptations.[108]

Just as a 'sneak' or 'transvestite' male sunfish is still pursuing an essentially male sexual strategy, albeit by means of deceit and subterfuge, the comparison suggests that homosexual men are acting analogously and that what we call 'homosexuality' in this respect might in fact be a closely comparable form of cryptic sexuality. In other words, so-called 'homosexuality' could be seen – at least from the evolutionary point of view – as a deceptive, subordinate sexual strategy adapted to promote the ultimate reproductive success of its possessors just as surely as any other. Its defining characteristic would not be its apparent sexual object –

---

[107] Dominey, 'Female Mimicry in Male Bluegill Sunfish', p. 547.
[108] See above pp. 39–40

members of the same sex – but its use of deception as a kind of behavioural camouflage behind which to hide its true fitness-maximizing purpose. From this point of view what we know as homosexuality would only be one – albeit one very important – example of the wider biological phenomenon of deceptive or cryptic sexuality.

Furthermore, it would need to be distinguished from the extreme 'transsexualism' discussed earlier, because the latter does indeed seem to presuppose a non-reproductive role for the trans-sexual in question. Up until now, sociobiological thinking about homosexuality has tended to assume that all homosexuals are non-reproductives, and, to the extent that this is true, I am in agreement that purely non-reproductive homosexuals should be seen essentially as 'helpers in the nest'. As with sterile castes of workers in insect societies, the genes which determine their non-reproductive lives might be transmitted preferentially by their parents and aided by their 'altruistic' dedication to their close reproductive relatives.

However, I cannot believe that the whole of human homosexuality can be seen in this way, especially if one recalls the facts relating to the spread of AIDS through the male homosexual population and the astonishing promiscuity and hypersexuality which surveys of its behaviour have revealed. What we are terming 'transsexualism' probably can be interpreted in terms of the sociobiological theory above, but that may be an extreme form of homosexuality, if indeed it is one at all. In more general terms, it seems possible that – 'transsexualism' apart – homosexuality should be understood as a form of cryptic sexuality: that is, one in which males pursue their reproductive aims covertly, often under pretence of appearing to be non-male, possibly even female. Yet, as I pointed out in my earlier, general discussion of cryptic sexuality, they remain male nevertheless.

In the light of this consideration it is not surprising that

> sex differences in the behavior of homosexuals ... seem to match or exaggerate those in heterosexuals. The men are ardent consumers of pornography, whereas there is no such market among women. Men judge the attractiveness of potential partners by their physical beauty, and especially by their youth, to a far greater extent than do women.[109]

---

[109] Daly and Wilson, *Sex, Evolution and Behavior*, pp. 308–9.

This finding is corroborated in Freud's *Three Essays* where he comments that

> There can be no doubt that a large proportion of male inverts retain the mental quality of masculinity, that they possess relatively few of the secondary characters of the opposite sex and that what they look for in their sexual object are in fact feminine mental traits. If this were not so, how would it be possible to explain the fact that male prostitutes who offer themselves to inverts ... imitate women in all the externals of their clothing and behaviour?[110]

If we regard apparent homosexual behaviour as deceptive, but fundamentally masculine, we should expect to find fairly close behavioural approximations to the two principal physical types exemplified in the evolution of cryptic masculinity: the immature, or *paedomorphic* (literally, 'child-like') form which avoids competition by appearing to be too young to be a threat to other mature males, and the *feminine* or 'transvestite' form which appears to be a member of the opposite sex. Although Freud only mentions the latter in the quotation above, he might have added that, transvestites apart, the other most popular love-object for the male homosexual is the boyish, or paedomorphic type.

This appears to leave us completely in the dark about why, apart from their notorious indiscriminateness, other males should find these deviant males at all interesting as sexual objects. Surely, males who satisfy their sexual drives with such boyish or feminine substitutes are being deceived – something for which evolution would certainly not reward them?

However that may be, an important finding of psychoanalysis suggests that the whole picture does make complete and compelling sense. Analysts habitually find that what attracts a man to another male, be he boyish or feminine in appearance, in preference to a female is the male's possession of the penis. This, as we may now expect, is an aspect of the castration complex. Evidently, absence of the penis in the love-object creates anxiety which contributes to the formation of the alternative sexual strategies in various ways. From this we must deduce that a male who cannot tolerate a love-object which might be seen as castrated has failed to pass the test of the Oedipal period and adopted a subordinate, deceptive sexual strategy

---

[110] Freud, *Three Essays*, VII, 144.

for adult life. Such males may come in at least three basic types, although the range of possible sub-types seems almost limitless.

First, we can envisage a male who, while confident enough of his own masculinity to adopt a masculine attitude as a sexual *subject*, has still failed to master the infantile castration complex sufficiently to feel able to compete with other males for women as sexual *objects*. Instead, he adopts love-objects which arouse the castration complex less by being endowed with the penis and for which he does not have to compete with 'regular', less ambivalent males. His objects will be the next two types.

Secondly, we can envisage a male whose Oedipal resolution is even less reassuring and who cannot adopt a masculine attitude either as a sexual subject or with regard to an object. His Oedipal resolution is 'negative' in the sense that he lacks an adequate masculine identification from his father and has instead tended to identify with women, so that subjectively he feels 'feminine' rather than masculine. He is a homosexual of the 'transvestite' type and an obvious object for those of the preceding type.

Thirdly, we could envisage a male with an essentially unresolved Oedipal stage who remains 'immature' in the sense of never adopting any particular Oedipal resolution, masculine or feminine, as the case may be. Such a male is the psychological equivalent of the immature, paedomorphic form which avoids confrontation with regular males by never reaching a mature, sexually dimorphic appearance – at least from the behavioural point of view. Of course, many young boys may in fact be in this condition when they attract the attention of older homosexuals of the first type and, clearly, homosexuality and paedophilia are closely related.

While these may represent the classical forms, many other possibilities certainly exist. For instance, males who adopt themselves as their own love-objects may constitute a special case of the first type. This could be seen as an evolutionary, adaptive basis for *narcissism*, at least as a sexual perversion, and for masturbation, at least in so far as it can be seen as *auto-erotic*.

According to the findings of psychoanalysis, *fetishism* also relates to the castration complex via the finding that in the unconscious the fetish represents the missing female phallus. We have already seen that it has precedents in the behaviour of captive primates who are also clearly capable of adopting substitutes for the female genitals. In the case reported, it was a rubber boot – a common type of fetish among human beings too. [111] However, the observation that fetishism is an almost exclusively male perversion and one found in association with all of the preceding types of homosexuality as well as with heterosexuality is what should be expected if fetishism is interpreted as a defence against castration anxiety. It seems that the fetishist overcomes the anxiety provoked by the absence of the female phallus by means of a substitute which, if it is largely successful, allows him to choose a conventional, female love-object, but which, if it partly fails, directs him to object-choices similar to those of the other types.

Such an approach as this might also explain homosexuality in women, which, according to the *Three Essays*, 'is less ambiguous [than it is among men]; for among them the active inverts exhibit masculine characteristics, both physical and mental, with peculiar frequency and look for femininity in their sexual objects.'[112]

The starting point for an explanation of female homosexuality has already been suggested earlier when I implied that exploitation of sexuality in infancy to secure enhanced parental investment may have produced conditions in which penis-envy might actually motivate a little girl to behave like a boy and assume masculine character traits, even though she remained physically female. The notable development of penis-envy in female homosexuals reported by analysts supports the contention that lesbianism may be an extension of it to the point where it motivates pseudo-male behaviour which, once successfully established in childhood, becomes a norm for adult life.

Furthermore, and as we noticed earlier, the asymmetry of the sexes with regard to sexual selection means that, whereas males are powerfully moulded by female choice, females are likely to be less affected by male choice. Presumably this means that they have greater freedom to vary their behaviour than do males, given that most females in a polygynous species end up mated, but only some males do, or do so only some of the time.

Lesbianism may have positive adaptive value in a polygynous species where a number of perhaps unrelated females will have to

[111] See above, p. 33.
[112] Freud, *Three Essays*, VII, 154.

try to live together in one domestic group. Pseudo-masculine behaviour, perhaps on the part of a senior wife whose reproductive life is coming to an end, could actually enhance her ultimate reproductive success (perhaps by way of its positive effects on her offspring) and reduce conflict with other, younger wives, still reproductively active. For younger wives, tolerance of, or even actual attraction to, such masculine females could also pay if it reduced domestic conflict and consolidated their personal standing, perhaps especially if they were almost as dependent on their relationship with the senior wife as on that with the resident male.

Again, women who lack adequate male support or investment in themselves and their offspring may find some mutual assistance from one another and perhaps even form quasi-monogamous or pseudo-polygamous unions. Essentially, such behaviour would not be very different in its adaptive basis from that of the occasional lesbianism reported among five species of gulls. Such birds are monogamous, which means that normally a male and a female must cooperate together to raise their common offspring. Since unpaired females have little chance of having any offspring if they try to raise them alone, similarly placed individuals might team up to raise their joint progeny.[113] In effect, the female gulls in question would be cooperating in their mutual reproductive interest, much the same motive as I am attributing to human lesbianism.

If this were true, then homosexuality in both sexes would be fundamentally the same: a question of the suppression of dimorphic, regular sexual behaviour, and perhaps even the appearance of belonging to the opposite sex so as to reduce conflict or to create positive ties with members of the same sex. The simpler situation and generally less extreme and less common effects of this in the case of females would presumably reflect the fundamental fact that among them sexual conflict is typically much less acute than it is among males. Again, the fact that female parental effort is normally greater than female mating effort suggests that women have less incentive to adopt cryptic forms of sexuality than do males, where the reverse is true.

The expectation that homosexuality in either sex should be some kind of abnormal aberration may well derive from precisely the same kind of misapprehension about the sex ratio as I mentioned

[113] Trivers, *Social Evolution*, p. 199.

earlier.[114] As long as one assumes that a one-to-one sex ratio indicates that monogamous mating is 'natural', any other kind of arrangement, like polygamy, might seem bizarre. More bizarre still might alternative strategies like homosexuality seem, yet, since polygamy creates a significant number of unmated males in a species with a unitary sex ratio, homosexuality could in reality seem as natural as anything else, at least among males.

Although males pursuing an apparently 'deviant', homosexual strategy in primal societies would not be expected to behave like 'regular' males and become prominent polygynists, they might nevertheless perpetuate the genes for their own behaviour thanks to covert copulations with otherwise 'regularly' mated females. Precisely the same argument applies to female homosexuals, whose emotional ties with other women in their domestic unit would not necessarily preclude occasional impregnations by males outside it. Unfortunately, reliable data regarding the heterosexual contacts of homosexuals of both sexes in modern societies are hard to come by, and may well give a false impression of primal conditions even if they were available. Nevertheless, the impression one gets from case-histories and those accounts that are available suggests that this is an important, if often overlooked, dimension.

Certainly, if this theory of homosexuality and other sexual deviations is correct, we might begin to suspect that, in general, homosexuals in particular and sexual deviants in general might be found to possess more latent heterosexuality than might otherwise have been imagined. For instance, figures quoted in one recent American study suggest that up to 50 per cent of homosexuals do indeed have children.[115] Admittedly, the same study shows, as one would expect, that heterosexuals have more children; but the point is that homosexuality may be a part of a *subordinate* sexual strategy for human males, like sneak fertilization or transvestism for bluegill sunfish. Sneaks and transvestites never father the majority of offspring, but they always father enough to keep their own particular sexual strategy going. Similarly, with human males, it may be that homosexuality can never become the dominant sexual strategy, but is always likely to remain an important subordinate one. Only if we entertain

[114] See above, pp. 51–3.
[115] M. Ruse, *Homosexuality*, p. 142, quoting the Second Kinsey Study.

out-dated Social Darwinist notions about sex being for the benefit of the species rather than for that of individuals does a proposition like this seem 'unnatural' or counter to evolutionary logic. Once one takes the individual's point of view into account, subordinate or apparently 'deviant' sexual strategies can be accounted for and are to be expected.

As far as Freud's investigations of male homosexuals were concerned, he evidently found that

> in individual cases direct observations have been able to show us that the man who gives the appearance of being susceptible only to the charms of men is in fact attracted by women in the same way as a normal man; but on each occasion he hastens to transfer the excitation he has received from women to a male object, and in this manner he repeats over and over again the mechanism by which he acquired his homosexuality.'[116]

One further reason for thinking that homosexuality should be regarded as a case of cryptic sexuality rather than the non-reproductive adaptation which it appears to be is the fact that homosexuals of either sex do not seem to be an identifiable sub-group in primal societies in quite the same way in which they have emerged in more recent cultures. Among Australian aborigines there is certainly little evidence to be found of men or women who consciously represent themselves as homosexuals or are identified as such by their peers. On the contrary, the fact that most women are married in polygynous marriage systems means that, among primal hunter–gatherers in Australia, few women would be in a position to practise self-conscious, overt lesbianism, even if they wished to do so.

As far as men are concerned, much the same seems to be true: homosexuals are simply not defined as such. Indeed, a colleague tells me that in one of the aboriginal groups he studied no word for homosexuality existed prior to contact with whites, so that the English word 'puff' had to be imported to describe a reality which appeared to need no term of its own beforehand.[117] Such reports are common. Even in the case of the broad-minded and rather

[116] Freud, *Leonardo da Vinci*, XI, 100.
[117] Personal communication from Dr David McKnight. The group in question is the Wikmungan.

Freudian Mehinaku, there was no word for homosexuality in the language and 'the very idea of homosexual relations' was considered ridiculous. But, significantly for the view propounded here, the one man in the past who might have been a homosexual was called 'Slightly a Woman'.[118]

Of course, this does not mean that latent homosexual proclivities exist in aboriginal human beings to any less a degree than they do among others. On the contrary, the fact that, as Freud himself declared, 'we ... cannot reject the part played by unknown constitutional factors'[119] which are presumably present in all human populations adds further weight to the realization that, whether manifested as a self-conscious social role or not, homosexuality is likely to be a factor in all human groups, aboriginal ones included.

Furthermore, one should recall that, at least as far as men are concerned, conditions of primal polygyny dictate that at least a few males will never have official wives, and many will only have one. For less successful polygynists such as these, the benefits of latently homosexual, submissive and possibly even pseudo-feminine behaviour might be considerable, at least on occasions. If those occasions were especially crucial to their ultimate reproductive success because mainly concerned with sexual opportunities, the behavioural genetic determinants of what we call homosexuality could well become established as a subordinate mating strategy in primal conditions.

Again, the fact that a homosexual social identity does not seem to have existed in primal societies should remind us of an important observation which also argues strongly for the cryptic theory of homosexuality. Up until now I have spoken as if homosexuality and 'regular' masculine or feminine behaviour were distinct types, wholly identifiable with individual persons. Sometimes – but apparently only in post-primal societies – this seems to be the case. However, in general it would be wrong to give the impression that cryptic forms of sexuality are all-or-nothing affairs.

On the contrary, analytic investigations confirm what common observation suggests: that constitutional human bisexuality ensures the presence of both masculine and feminine, regular and cryptic forms of sexuality in one and the same person. No Oedipal resolu-

---

[118] Gregor, *Anxious Pleasures*, pp. 59–61.
[119] Freud, *Leonardo da Vinci*, XI, 101.

tion is ever so clear or so definitive that it excludes its opposite, and no degree of identification with the same-sex parent entirely precludes a latent tendency to internalize the other. Freud frequently drew attention to the fact that heterosexuality and homosexuality normally co-exist in varying quantities in the sexual constitution of everyone so that 'in addition to their manifest heterosexuality, a very considerable measure of latent or unconscious homosexuality can be detected in all normal people.'[120]

Finally, the possibility must be considered that the prime adaptive value of homosexuality was much greater in the past than it is today, even among primal hunter–gatherers such as the Australian aborigines. I shall present considerations below which will suggest that this may well be the case and that, whatever possible adaptive value homosexuality may have in recent and modern societies, it may have had vastly more in the distant past when primal hunter–gatherers originally evolved.

[120] Freud, 'A Case of Homosexuality in a Woman', XVIII, 171.

3

Female Choice in Human Evolution

Brain, brawn and hunting

I ended the previous discussion with the suggestion that human beings are notable because of the way in which they manifest cryptic sexuality, not as external, somatic adaptations as most other organisms who exhibit it do, but as purely internal, psychological tendencies. The time has now come to consider how this could have come about, and why evolution should have taken this extraordinary turn in the human case. As we shall see shortly, these considerations give a most unexpected but welcome bonus in providing a startling new answer, not merely to the specific question outlined here, but to the more general and fundamental one concerning human evolution as a whole and, in particular, the problem of explaining the apparently inexplicable expansion of the organ on which human psychology depends – the brain.

In considering the matter of the evolutionary causes of the development of the human brain it is easy to succumb to fallacious thinking. With the benefit of evolutionary hindsight, the evolution of the human mind seems self-explanatory. After all, how else would we have got where we are today? But it is completely fallacious to believe that the human brain evolved 'in order' to make culture and civilization of the kind we have today possible. While it may be perfectly correct to comment in retrospect that this is what it did enable, it would be putting the evolutionary cart in front of the adaptive horse to argue the other way around.

Although the big-brain-evolved-to-make-modern-technology-possible theory is evidently false, a restricted version of it seems to

142

have gained wide credence, despite being just as spurious. This is the theory which states that the modern human mind and brain evolved to facilitate *hunting*. As we shall see shortly, this theory may be true in a highly modified form, but as a simple and straight-forward explanation it will hardly do.

The argument that evolution endowed our hominid ancestors with intelligence 'in order' to adapt them for hunting only shifts the question from one aspect of the problem to another. For if this were so, we are still ignorant of why evolution should have endowed us with unique intellectual adaptations relating to hunting rather than the more common mammalian hunting adaptations rep-resented by claws, fangs and fleetness of foot.

Specifically, we must confront the question why hominids with big brains 'for' hunting would have had greater reproductive success than other hominids without the brains but with the brawn. Indeed, the existence of some very brawny hominids in the past suggests that such a trend in evolution would appear to be very much more obvious than what we think we see: less brawny, more brainy hominids, apparently adapting to hunting by developing enormously enlarged intellectual capacities, but retaining the digestive system and many other adaptations of a more or less full-time vegetarian forager.

Finally, the argument that human beings are particularly intelli-gent because they had to 'invent' hunting skills and learn to use tools loses much of its force when we notice how extensive tool use is among other animals, especially our near primate relatives, the chimpanzees. Manipulating sticks and stones because we lacked hunters' teeth and claws may require some measure of cerebral development, but I can hardly believe that it explains the genius of an Einstein or a Leonardo da Vinci, especially if we notice that the sum total of human tool-making skill consisted of chipping stones, carving sticks and shaping bones until just fifteen-thousand-odd years ago. Furthermore, whereas tool technologies advance in stages, with long intervals between advances, brain growth appears to have been a continuous process and one that cannot be closely tied to the evolution of particular stone technologies as such.

Nevertheless, the fact remains that hominid brain growth and the beginnings of human hunting do seem to go together, suggesting that, even if a simple cause-and-effect relationship does not exist, a more subtle and indirect one might. Finally, it is worth pointing out that since there is usually a sharp division of labour along sexual lines in hunting and gathering societies, an explanation linking hominid brain growth to what was almost certainly an almost exclusively male undertaking leaves the explanation of female brain growth – by no means less spectacular than that of the male – entirely out of account.

Matings for meat

However that may be, consideration of the foregoing arguments suggests a much more credible theory of human evolution;[1] one which, far from leaving one sex entirely out of account, accords it a crucial role in exercising a unique privilege of choice.

As we have seen, evidence of considerable sexual dimorphism, along with that relating to the size of the testes, would be enough to exclude both monogamy (contradicted by the former) and promiscuity (ruled out by the latter) as credible evolutionary foundations for human mating, leaving only polygyny as a likely candidate. When we add to this the arguments already advanced relating to the Australian aborigines and to probable evolutionary parallelism with the gelada baboon, we have a very strong case indeed that polygyny, the most common form of mating in recent human societies, may indeed have been characteristic of the hominid ancestors of modern human beings.

But this immediately creates a difficulty with regard to hunting, because one-male group polygyny of the kind evidenced, for instance by the gelada baboon, seems a very unpromising social structure for a cooperative hunter to adopt. Essentially, the theory advanced here holds that it was a unique combination of one-male group polygyny and multi-male cooperative hunting which produced, as an indirect but fundamental consequence, the anomalous expansion of the human brain and the whole phenomenon of Oedipal behaviour, cryptic sexuality and dynamic human psychology discussed earlier. Furthermore, the theory states that, as Darwin seems to have intuited in entitling his great work on human evolution, *The Descent of Man and Evolution in Relation to Sex*, female choice was the

---

[1] Partly anticipated by D. Symons, *The Evolution of Human Sexuality*, pp. 139–41, J. H. Crook, *The Evolution of Human Consciousness* and K. Hill, 'Hunting and Human Evolution'.

crucial factor in bringing about these most characteristic of human adaptations.

In order to understand how female choice may have come to mould the evolution of men, let us return to the considerations which we left at the end of the previous discussion. We know that our hominid ancestors began to hunt, presumably after moving out onto the savanna grasslands and perhaps acquiring their gelada-like adaptations. Besides the grains and tubers which they probably exploited as foragers analogous to the gelada baboon of today, they would have found huge herds of animals which at some point we know they began to hunt.

We also know that the hunters were overwhelmingly likely to have been males, rather than females, and young males rather than older ones. This is because pre-existing sexual dimorphism may have made males more adapted to hunting than females in the sense that they would have been physically bigger and carried less fat and rather more muscle.[2] Women, by contrast, would have been most unlikely to have the freedom to undertake extensive chases either because of the burden of young – born or unborn – or because they could not escape the control of their harem-owner. The greater aggressiveness of males and their known likelihood to undertake greater risks and to leave their natal units would, by contrast, have made them much more likely to be the first hunters. Finally, if we look for precedents in closely related or analogous species, 'both the chimpanzee and baboon data indicate that hunting is almost strictly a male activity.'[3]

But more than anything else, the association of young males in all-male bands, unencumbered with females and young, would have made this social structure ideally pre-adapted for cooperative hunting. This is because individual hunting would have been dangerous, to say the least, for hominids unequipped with natural weapons and defences in the form of teeth, claws or great turn of speed, and as yet unprovided with the hunting technology which culture would eventually produce. An unequipped, individual, day- or night-time hunter would have been more likely to be a victim of much better adapted competitors than to succeed in establishing a new mode of subsistence by himself.

[2] See above p. 56.
[3] Hill, 'Hunting and Human Evolution', p. 533.

But social hunting would have been quite a different matter. While an individual hominid would have been highly vulnerable, an entire group would have been much less so. Furthermore, studies of cooperative hunting in a number of other mammalian species show that both the size of prey brought down and the success-rate of the hunt increase dramatically with the size of the group involved. This would almost certainly have been an even more important factor for early hominid hunters, unequipped and maladapted as they would have been by comparison with other hunting mammals.

This would have been true even if human hunting began with scavenging carcasses already on the ground rather than active hunting of live prey.[4] This is because even in the case of scavenging a considerable group would be necessary to take over a carcass from what could well be formidable opposition from hyenas, vultures, lions and leopards. And of course, even a successful hunt would involve defending the kill from such competitors once it was brought down.

If we look at the beginnings of hunting from the point of view of cost and benefit to the individual male, rather than from the fallacious perspective of the group, we immediately notice an important factor derived directly from the considerations set out earlier. This is the fact that, as we saw, because the sex ratio tends to remain near unity even in a polygynous species, both the reproductive success and the vulnerability of males tend to be much greater than those of females. One consequence of this is that highly dangerous activities which might confer a reproductive advantage become attractive to males in a way that they never would to females of the species, because female reproductive success is more reliable, and consequently does not attract risk-taking. Here, hunting may have been a case in point.

If we assume that cooperative hunting was a possibility, whereas individual hunting was not, then males may have cooperated to undertake it in much the same way as other herding behaviour is motivated – by mutual self-interest. This is because a group can provide a convenient shelter for the individual, and one can hide behind the other just as easily as the other can hide behind one. In this way other animals can be used as barriers and hiding places by individuals, who can minimize their personal exposure to predators

---

[4] F. Szalay, 'Hunting-scavenging Proto-hominids: A Model for Human Origins'.

by this means. If all members of a local population pursue this tactic, large aggregations will form, with individuals exploiting the personal advantages to themselves of coordination in respect of their spatial distribution.

No kind of 'herd instinct' or 'social whole' need be imputed in these cases, and there is no justification for thinking that anything other than individual self-interest is the primary motive for most forms of herding or shoaling behaviour. In these cases it is seldom a question of individuals paying a price for group membership which secures benefits for the group as a whole. On the contrary, it is rather one of the benefits from belonging to the group accruing more or less equally to all, not as members of some social unit greater than the sum of its parts, but as individuals who exploit the group for their personal benefit. This logic applies to groups which hunt just as well as it does to those which are hunted; and all the more so if, as we have already seen, hominid hunting groups were exposed to the danger of being hunted by other, better adapted hunters with whom they were competing, such as leopards, lions and hyenas.

Cooperative hunting, however, was still probably quite dangerous for any particular individual, involving as it would have done lengthy chases and dangerous confrontations both with the animals hunted and with other hunters and scavengers. Nevertheless, if the ultimate reproductive benefit was sufficiently great, exposure to danger may have been well worth the cost to any individual, especially if it evolved in conditions of polygynous mating where only a minority of males could expect to command harems at any one time and where a majority found that they had both the opportunity to hunt and a ready-made social unit in which to do it.

Furthermore, the possession of meat would almost certainly have been a highly attractive bargaining counter in interaction with females. If we assume a chimpanzee-style social structure of early hominids, with females pairing off with particular males during their most fertile periods, then it is easy to understand how a meat-possessing male might attract a female and obtain copulations as a result.

But even if we consider the apparently less promising gelada-style social structure, the possession of meat could have made a great difference. I call this situation less promising because it seems as if a hunting male's problems are much greater and his chances of success much more slender in conditions where - as in the modern gelada case – all females are herded into harems by dominant males who hold competitors at bay. This would appear to suggest that conflict among males is the deciding factor in securing access to fertile females, and certainly seems to be decisive in serious attempts at the takeover of a harem.

However, field studies of such takeovers show that, although, as expected, conflict among males is intense, 'the actual process that brings about victory or defeat is much more complex and far from being a case of brute force and ignorance.'[5] On the contrary, 'the outcome is decided either by the intruder giving up and returning to his all-male group or by a female interacting with him.'[6] In other words, although not being defeated by a resident male is a necessary condition for harem takeover, it is not a sufficient one. For a takeover to be successful, an intruding male must also be accepted by at least one of the females (and in that circumstance can usually expect others to follow quickly). In short, female choice is just as important as inter-male conflict.

This is also true if a male pursues the second, subordinate tactic, namely joining an existing one-male group as a 'follower'. In order to do this he needs to avoid conflict and confrontation with the dominant male, but in this case, even more than in the preceding one, establishing relationships with as many females as possible is vital if his membership is to become permanent, so that, here again, female choice is extremely important.[7]

Even if the first hominid hunters were indeed members of all-male groups set within the matrix of a wider society featuring monopolization of females by single males, it is possible that possession of meat could have been a crucial factor in influencing male reproductive success, via the phenomenon of female choice. A meat-bearing male might have been especially attractive to a hominid female if, like today's gelada baboon, she had otherwise been a bulk feeder on a low quality food source, spending 50–75 per cent of the time foraging.[8]

Data summarized by Hill in a paper which anticipated the point I am making here show that nutritional returns from vegetable items

[5] Dunbar, *Reproductive Decisions*, p. 130.
[6] Ibid., p. 132.
[7] Ibid., p. 143.
[8] Ibid., p. 115.

can be expected to fall into a range from 128 to 5,984 calorie-hours whereas meat items, by contrast, range from 1,215 to 65,000 calorie-hours.[9] But this disparity is in its turn overshadowed by what appears if we take into account the hunter's ability to carry food items. Hill shows that someone with no carrying implements returning some distance might be able to carry about 800 calories of fruit or vegetables, but could easily manage 39,000 calories-worth of meat.[10] Furthermore, the expectation that females would prefer meat if they could get it is confirmed by field studies of baboons and chimpanzees, which show that they do indeed have a taste for raw meat, and will greedily consume it if given the chance. If meat were only obtainable by the cooperative hunting ventures of all-male groups, the young males who predominantly constituted such groups would have possessed a most desirable advantage in trying to secure access to the females of one-male foraging groups, otherwise unable to obtain the valuable new food resource.

One good reason for thinking this is the fact that similar events can still be discerned in some hunting societies. In the case of the Sharanahua of South America,

> as for other tropical forest Indians the incentive for hunting is to gain access to women, either as wives or as mistresses. One can see variations on a single theme from the crude gift of meat 'to seduce a potential wife' among the Sirionó; the elaboration of the special hunt among the Sharanahua; to the young Shavante's provisioning his father-in-law with game after the consummation of his marriage ... Whether men prove their virility by hunting and thus gain more wives or offer meat to seduce a woman, the theme is the exchange of meat for sex.[11]

This is unmistakable in the case of the Sharanahua 'special hunts' just mentioned:

> The special hunt is started by women ... Each woman chooses a man to hunt for her, a man who is not her husband nor of her kin group, though he may be her cross-cousin, her husband's brother, or a stranger.

[9] Hill, 'Hunting and Human Evolution', p. 534. A 'calorie-hour' is defined as 'calories divided by time necessary to extract and process the item once it has been encountered.'
[10] Ibid., p. 536.
[11] J. Siskind, 'Tropical Forest Hunters and the Economy of Sex'.

The connection I am suggesting between the provision of meat and the role of the male became so unmistakable to one indian wife that 'when the men of her household returned from the hunt, tired, depressed and empty-handed,' she was heard to remark, 'There is no meat, let's eat penises!'[12]

Returning to what may be the evolutionary origins of these hunts, we might suppose that at first particular young males could have obtained meat in cooperative hunts in order to use it as a bargaining counter with females, either following violent confrontations with their existing owner, or after having been accepted as his follower. Indeed, in the latter situation it is conceivable that a follower, having obtained meat from a cooperative hunt with an all-male group, could also use it to placate the dominant male, so that meat became a factor not merely in female choice, but in interactions among males as well.

We have already seen that we must presume that females were debarred from hunting: by being encumbered with young, by adverse sexual dimorphism (as far as hunting was concerned) and by the need to be guarded, or at least, closely supervised by males for whom, even in a hunting economy, they remained a scarce and critical resource. Yet at the same time, the provision of meat would have been especially critical for females, characteristically concerned as they are with parental, rather than mating, effort.

Although hunting males' confidence in their own paternity of offspring born to such females would probably have been low, given that we are assuming that the latter might have remained in harem groups dominated by older, non-hunting males, the fact remains that hunting males could not prevent females provisioning their offspring with some of the meat, if that was their wish. In any event, better fed females would be likely to have more regular sexual cycles and presumably would also conceive and gestate more reliably as well. This would mean that, confident of paternity or not, males provisioning females in the circumstances we are envisaging could not divorce such provisioning from those females' direct or indirect investment in their offspring, whoever the father might have been.

After all, the choice for the otherwise unmated hunting male lay between a chance of having offspring by means of matings traded for meat and having virtually no chance of mating at all.

[12] Ibid., pp. 234 and 233.

Since something is better than nothing, the possibility that such males' gifts of food were used as investment in the offspring of other males was a factor which they would have to accept, the alternative being worse. Indeed, given that chance would dictate that females in a dominant male's harem group would conceive by him on some occasions, provisioning and thereby investment in his offspring by hunters would clearly have been in his reproductive interest, and perhaps provide another reason why such males might tend to tolerate at least some extra-mural matings by their harem-members. If something like this did begin to occur, human evolution would have taken a decisive turn which was to have far-reaching effects on both males and females, offspring and parents.

Cryptic estrus

If we consider the female of the species first, the evolutionary scenario sketched out above may explain one of the most notable features of the sexual physiology of modern women: the absence of visible estrus, or, in other words, external signs that a woman is about to ovulate.

In many primates the period known as estrus is easy to identify because it is advertised by the female. In the case of chimpanzees, this advertisement takes the form of a large swelling around the female genitals which male chimpanzees evidently find fascinating, and which develops as the female in question enters into that period of her menstrual cycle when she is most likely to conceive. In many other primates comparable displays are found and in the intriguing case of the gelada baboon – the only primate apart from human beings to show a secondary sexual display on the chest of the female, it will be recalled – the hairless area in question changes colour while the polyp-like vesicles which adorn it undergo changes in size and appearance, all nicely synchronized with the onset of the female's most fertile period.

At first sight, such signalling of receptivity seems almost altruistic and one can well imagine believers in group-selection imagining that females of these species display their sexual state in order to promote efficient reproduction of the species, with minimum 'wasted' copulations and so on. In fact, one does not need to make any such assumption. If one looks at the polygynous mating system from the point of view of any one female, we can see that it is in her selfish interest to secure the attentions of the dominant male when she is most likely to conceive successfully by him.

This is easy to understand in the case of a primate like the gelada baboon, where females are dragooned into one-male groups dominated by a single, paternal male. Since such a male has a number of females to service, any one female who is particularly likely to conceive by him is best advised to advertise the fact if she wishes to maximize her eventual reproductive success. Consequently, gelada males, for instance, waste little time courting or copulating with their females outside the estrus period and when copulation does occur it is a fairly routine, perfunctory affair.[13] As we have already seen, in the case of chimpanzees, estrus is advertised, and sexually receptive females tend to consort with a single male who attempts to repel other potential mates during her fertile period.

If we now compare the modern human female with her chimpanzee or gelada equivalent we notice a striking difference. By contrast to the vast majority of primates, human females do not show any evidence of a fertile period; on the contrary, there are good reasons for thinking that, far from being advertised, estrus in the human female is actively *concealed*. There is certainly no obvious, external signal comparable to that seen in chimpanzees or geladas: the perineal area of the human female does not undergo any obvious enlargement, and although slight changes in the size of the breasts may occur, women do not show the conspicuous changes in shape and colouration seen in the chest ornamentation of the female gelada. Admittedly, there is some evidence of a heightening of sexual desire among human females during the middle, most fertile part of the sexual cycle; but this behaviour, if real, is subtle by comparison to the sexual mania which grips most female primates during the estrus period. 'It is a truly remarkable attribute of human females that their ovulation is often essentially impossible to detect, even, in some cases, through medical technology.'[14]

A possible explanation for this concealment of estrus relates to the foregoing scenario of the origins of hominid hunting. But

[13] Dunbar, *Reproductive Decisions*, p. 52.
[14] R. D. Alexander and K. M. Noonan, 'Concealment of Ovulation, Parental Care and Human Social Evolution', p. 442; N. Burley, 'The Evolution of Concealed Ovulation'; L. Benshoof and R. Thornhill, 'The Evolution of Monogamy and Concealed Ovulation in Humans'.

before we look at that let us first consider an alternative explanation which may occur to some readers.

Following my earlier argument some may well have been tempted to reply to these considerations as follows:

'Human females avoid advertising estrus, as you call it, because they are not apes or monkeys and do not want to be subservient to so-called "dominant" males. Indeed, if we take seriously what you said earlier about looking at costs and benefits in analysing sexual behaviour, we can see that there is a flaw in your argument about why it might pay a female to advertise estrus.

The flaw is this: females have to advertise – or so you say – because they are polygynously mated: several females to one male. You would probably argue that this is in their interests because the male they are mated to has been called by you "dominant". They mate with him because he can drive off rivals. That is all very well, but it ignores the hidden cost to the females concerned when we consider their investment in their own sons.

Because females allow themselves to be herded into harems by males, their own sons must face competition for mates so that, in the hypothetical case you set out earlier, only one in ten males might actually mate at any one time. This means that a female who obligingly helps a dominant male to be a polygynist by advertising her fertile period also hinders her sons – or, at least, about nine out of ten of them – from having offspring. Since you are so insistent about evolution being all about eventual reproductive success, you must admit that, taking the costs of polygyny for their own sons and the benefits for themselves into account, females lose out.

It would be far better for the females in question to do what women actually do and not advertise their fertile period. This would make it more difficult for males to monopolize large numbers of females and would promote monogamy, along with the eventual reproductive success of all a female's sons, not just some of them. The only real reason why baboons are herded into harems is that they are not human. If they were they would not do it and your entire argument about polygyny being "natural" for human beings would collapse completely.'

Unfortunately, the assumptions on which this argument is based are completely fallacious. However, a careful consideration of them will clarify the logic of my argument and do much to illuminate the theory.

In part, this is because my imaginary protagonist ignores the point made in the first essay relating to the *relative* reproductive success of males. As we saw earlier, all other things being equal, the reproductive success of some males exactly compensates for the relative failure of others, so that a successful male offspring makes up for the loss involved in other males who do not succeed in breeding. On average, and ignoring the qualifications pointed out earlier, parents who have ten male offspring of whom only one might mate find that the reproductive success of the one exactly compensates for the failure of the other nine. Nevertheless, there is a more fundamental flaw which is best seen by imagining what would happen if the scenario sketched out in the objection outlined above were actually to come about.

Let us take the best possible case. Let us imagine that human beings had evolved early on into what chimpanzee society was at first thought to be like. Let us also suppose that females did not advertise estrus and that they refused to associate with any particular male for any considerable time, and refused to make any kind of prejudicial choice of mate. The mating system would then be a promiscuous one and could be imagined to have the desirable side-effect of promoting male cooperation with females and each other. No individual male could know for certain who his offspring were because he could not know when any female was most likely to conceive. Since females are assumed to be mating promiscuously with several males, all males would be in the same position so that all would have an incentive to cooperate in looking after the women and children. Sexual promiscuity would produce equality, cooperation and peace![15]

Such a picture sounds idyllic, but consider the following possibility. Imagine that some individual female fails to conform and instead of mating randomly and promiscuously 'for the good of all' instead does so for her own benefit, perhaps making a choice based on the sexual attractiveness or some other desirable characteristic of a particular male. Since her chosen mates will, by definition, be more desirable than those available by purely random mating, the

[15] According to statements made by Jane Goodall at a public lecture which the author attended in Detroit on 5 May 1988, this is actually what occurs among chimpanzees at the Gombe Stream Reserve.

male offspring of those mates would probably inherit their father's superior qualities, at least in the eyes of some females. If those offspring secure more matings than other males because of their superior attractiveness, their genes will spread through the population and, furthermore, so too will those of the 'choosy', discriminating females who are their mothers.

Before long female choice will begin to subvert the random mating order of promiscuity to the point where a female might prefer to be the second wife to an existing, mated male who had superior qualities in her eyes. Once having crossed this so-called *polygyny threshold*, the system could easily evolve into thorough-going polygyny, with reproductively most successful males being treated as a scarce resource by females and passing on more of their desirable attributes to their own male offspring.

'Group-marriage' of the kind implied in the promiscuous scenario above would only work if all its constituent individuals, both male and female, mated indiscriminately. But evolution does not reward such lack of choice, especially by females. Any particular female will always gain from the evolutionary point of view if she does discriminate in favour of the male who is, for instance, the better provider of parental investment, the possessor of slightly more appealing traits, the progenitor of marginally more, or more reproductively successful offspring.

The really fundamental fallacy in the objection quoted above is that it assumes that females *as a whole* could change a system based ultimately on *individual* reproductive advantage. There is no doubt that if females could cooperate on such a policy and, if none of them defected against it, promiscuity of that kind could evolve. The flaw in the argument is the assumption that they could indeed be relied upon to be completely indiscriminate where their self-interest was concerned.

This is a common error which goes to the very heart of social theory and which is based on the wishful thinking that whole groups could be made to cooperate in the presumed collective interest. Unfortunately, rigorous logical and mathematical analysis of such situations – not to mention bitter experience of life – proves that, in anything but the smallest groups, the cost to individuals of cooperation with the shared interest will usually exceed the benefit of defection in individuals' own self-interest. Conversely, a benefit shared by all in the group will not usually fully compensate for its added cost to all individual members.

In other words, 'free-riding' at others' expense will always be a real temptation and will usually introduce a sharp cleavage between individual self-interest and presumed collective benefit in all except the smallest and most intimate groups. Furthermore, even this qualification is far from universally true since, as we have already seen, even where two is company and three is a crowd, the differential costs and benefits to partners in this simplest and most fundamental group may induce real conflicts of interest between them.[16]

If we return now to the question of concealment of estrus, we can begin to see that, from the point of view of any individual female interested in being provisioned with meat by a male, concealment would be most desirable. This is because it would only promote the reproductive success of a male to exchange meat for copulations when a female was in estrus, since matings at other times would not result in offspring. Furthermore, as Hill has also pointed out, 'in both chimpanzees and baboons ... estrous females receive a greater percentage of meat from a kill made by males than do non-estrous females'.[17]

'But why should estrus matter to males?' I can imagine my antagonist protesting. 'After all, you have already argued that males are "naturally" less discriminate than females, sometimes to the point of mating with individuals of the wrong sex or species. If they are so lacking in discrimination about such vital factors as these, why should they bother themselves about a little detail like estrus?'

My reply is that, far from contradicting the fundamental principle which says that males should be less discriminate than females about sex, the situation which we are envisaging actually follows from it with an inexorable logic.

Let us look at it from the point of view of any particular male, who we presume is in competition with other males. If, all other things being equal, he 'wastes' meat on females who cannot conceive when he impregnates them, he leaves fewer offspring than a male who only gives meat in exchange for matings with advertising females. Since meat is in finite supply, along with estrus females,

---

[16] For a full discussion of this fundamental issue see Badcock, *The Problem of Altruism*, Introduction and essays 1 & 3.

[17] Hill, 'Hunting and Human Evolution', p. 533.

males who 'economize' in this way ought to leave more offspring than those who do not.

Investment by the male in the form of meat provided to females should produce some kind of discrimination on the basis of the principle that discrimination is proportional to the level of investment made by each sex and proper to the type of investment in question. In general terms Robert Trivers remarks that 'male investment implies male choice. There is evidence for this in several species ... Consistent with this, males discriminate on the same bases as do females, but are always less discriminating.'[18] In other words, a preference for mating with estrous females on the part of male hominid hunters in no way contradicts the basic predictions of the theory of parental investment, but vindicates them. Males, like females, will choose if and when it pays, even though in general terms they will almost always do so less than females because generally they will pay less than females.

The point becomes unmistakably clear if we consider the female's point of view because, as we have just seen, there need be no necessary cleavage of interests on this issue among females in a polygynous group if the latter receive only fertilizations from their male 'owner'. If only fertilization is at stake, both male and female have an interest in making sure that copulations occur predominantly during estrus. But if provisioning females with highly nutritious food is at issue the interests of the sexes differ dramatically.

From the point of view of a female, advertising estrus becomes a losing tactic if meat can only be had in exchange for copulations. This is because, as we have just seen, males who will give her meat when she is not likely to conceive will find themselves at a reproductive disadvantage compared to those who reserve their gifts for her fertile period. Consequently, as Hill has also pointed out,[19] females who do not advertise estrus would have a very considerable advantage in securing what is effectively enhanced paternal investment in themselves and their offspring via the provision of a highly desirable food resource. The fact that access to meat could only be had by males in the first instance because only males could undertake hunts inevitably meant that females had to develop adaptations of their own to exploit this valuable new resource. In this respect cryptic estrus might qualify as a prime female adaptation to hunting by males.

[18] Trivers, *Social Evolution*, pp. 255–6.
[19] Hill, 'Hunting and Human Evolution'.

The adaptive value of concealed estrus would therefore lie in its consequences for female diet and a woman's ability to feed her offspring with meat. Since we know that female fecundity is sensitive to body weight (which is itself a function of fat deposits), the general conclusion must be that females fed on meat might have a distinct reproductive advantage over those not so provided.[20]

Furthermore, these observations may not be without significance in the modern world because it is a widely observed fact that psychopathological eating disturbances affect women much more than men, and younger women much more than older ones. The sex-specific character of anorexia and bulimia may ultimately rest on precisely the evolutionary basis which we are discussing and might provide an evolutionary perspective to the link between female sexuality and eating which psychoanalytic investigations of these disorders has revealed. Later we shall find that we can suggest a fundamental cause for these disturbances which is entirely in harmony with our basic theme. For the time being, let us merely note this unexpected aspect of the problem and pass on.

But here I can imagine another very telling objection. 'Your theory is fine,' I hear my imaginary critic saying, 'but it could just as easily work the other way! Indeed, your reasoning might suggest that females had an evolutionary interest not in concealing estrus, *but in advertising it all the time.* This is all the more likely if, as we know to be the case, estrous females are more likely to be desirable to males with meat than others who are not in estrus.'

This is an excellent point, and may well be true! But it seems to me to come to the same thing in the end. The outcome must be that, by way of either permanent advertisement or total non-advertisement, females whose fertile period could not be divined by males would have an advantage by way of being more regularly provisioned and would become the dominant type. Furthermore, this is all the more likely because modern women do have another kind of permanent sexual display in the form of pubic hair, which first appears with the onset of nubility at puberty but fades again after its end at menopause.

But a permanent display of estrus could hardly be called a true periodic estrus display because it would always be apparent. Further-more, there is no evidence in modern human female sexual physio-

---

[20] R. E. Frisch, 'Fatness and Fertility'.

logy of permanent perineal swelling of the kind which distinguishes estrous chimpanzees. Admittedly, *steatopyga*, a prominent enlargement of the buttocks, is found in women of some African hunter-and-gatherer groups, and all women exhibit some degree of rounding out of that part of their bodies.

Nevertheless, if we look at the other primate which parallels human adaptations, the gelada baboon, we notice something more promising. As I mentioned earlier, geladas are the only primate apart from ourselves to exhibit a secondary sexual display on the female's chest. This comprises 'a chain of small, pink, fluid-filled beads or vesicles around the edge of each area of sexual skin, one of which is situated on the chest. The only exception occurs at puberty: *juveniles show a conspicuous swelling of the chest patch ...*' [21] Putting aside for a moment the interesting question of why it is only in the juvenile that this is seen in the gelada baboon, it seems there is indeed a parallel with human adaptations here. The breasts of modern women could certainly be described as a 'conspicuous swelling' of the same general area of the body and evidently do constitute a notable sexual signal. This suggests the possibility that, like young female geladas, female hominids of the distant past may have signalled estrus by a similar means. There is certainly evidence that in a large number of modern women the breasts can undergo enlargement by up to 20–25 per cent during sexual arousal, and nearly all women experience some slight changes in breast size and sensitivity during the menstrual cycle.[22]

Consequently, it is possible that what may once have been a periodic enlargement of breast tissue signalling ovulation has become a permanent adaptation in modern human females by way of being a *cryptic* form of estrus. In other words, it is one which may have begun as a reliable signal relating to a period of maximum fertility but which has perhaps undergone a modification by means of prolongation which has robbed it of that function and converted it instead into the opposite: a means of concealing the period of estrus by making it seem permanent.

Again, the observation that the cleavage between the prominent breasts of modern women seems to allude to that between the buttocks rather in the same way as the chest patch of the female gelada alludes to her perineal region suggests that the buttocks and

[21] Dunbar, *Reproductive Decisions*, p. 51, my emphasis.
[22] W. Masters and V. Johnson, *Human Sexual Response*, pp. 28–9 and 143.

breasts may to some extent have evolved their prominence together and that, gelada parallels notwithstanding, parallels with chimpanzee adaptations can also be seen.

At first sight such an adaptation as cryptic estrus would seem to relate purely to outwitting males. Denied reliable indications of exactly when a female was about to ovulate by the fact that she seemed always about to do so, males could not reserve gifts of meat for those critical times, but would be forced to widen their coverage and both mate and provide meat more frequently, to the evident advantage of the female in question. However, if the effect of this was continual monopolization of any particular desirable male or males by any particular estrus-concealing female, its effect might be to deny them to other females to the extent that they interacted with her rather than with others. In this way sexual competition among women becomes a factor relating to the evolution of cryptic estrus, along with inter-sexual exploitation of this particular form of deception.[23]

Menstrual synchronization

One further piece of evidence which seems to support the proposition that competition among females may have been an important factor in human evolution related to the origins of hunting is the fact of *menstrual synchronization* among cohabiting women. Scientific studies support the common observation that groups of nubile females who are in intimate contact with one another for extended periods of time will tend to find that their menstrual cycles are becoming synchronized to the point where, for instance, most members begin to menstruate on approximately the same day.[24]

At first sight this seems an impressive expression of natural feminine solidarity – what could be more 'sisterly' and cooperative than coordinating sexual cycles? However that may be, problems arise the moment one looks into the evolutionary background of

[23] To this extent, my theory closely follows that of Alexander and Noonan ('Concealment of Ovulation'). However, in certain respects it is the exact opposite because where they assume a multi-male setting and an essentially monogamous outcome, I envisage a one-male situation and a polygynous result.

[24] M. McClintock, 'Menstrual Synchrony and Suppression'; C. A. Graham & W. C. McGrew, 'Menstrual Synchrony in Female Undergraduates Living on a Coeducational Campus'; D. M. Quadagno et al., 'Influence of Male Social Contacts, Exercise and All-female Living Conditions on the Menstrual Cycle'.

such an adaptation. Nevertheless, first impressions seem more encouraging. This is because coordination is very common among many species of plants and animals and frequently or even characteristically applies to reproductive cycles. Many plants, for instance, time flowering to coincide with that of other members of their species; and in animal behaviour perhaps the most astonishing example of reproductive synchronization is a species of cicada, which remains underground but emerges to mate once every seventeen years! Again, female wildebeest who cross the African plains in herds of hundreds of thousands coordinate giving birth so that the vast majority of all one season's calves are born within a few days of one another. If ever there was evidence of group cohesion and evolution acting on the entire species, here it seems to be, because how else can one explain such impressive coordination of behaviour as this?

The answer is: very easily. Let us look at it from the point of view of any one individual female. Evolution is all about relative reproductive success, and wildebeest give birth, usually to a single calf, once a year. A new-born calf is highly vulnerable to predators who, if they catch it, will wipe out that particular female's entire reproductive success for that year. Yet all other females are in the same position, so that if any particular female times her giving birth to coincide with that of other females the likelihood is that her calf will only be one of a number available to predators. If all gravid females in the herd follow this strategy, all calves will be born within a few days of each other, swamping the local predators with available kills. But since the local predators can only eat a rather limited, finite amount of new-born wildebeest in any few days, the actual chances of any particular female losing her calf are minimized.

In other words, reproductive self-interest alone is enough to explain birth-synchronization among herding animals like wildebeest, and, as we have already seen, herding behaviour also admits of a similar interpretation. Ungulates herd and fish shoal because the chance of any one individual being seen or eaten by a predator is considerably less if that individual hides behind the others in the group (this is the behaviour which makes the herding of sheep by dogs possible). If all individuals are similarly motivated, all will coalesce, sometimes into enormous groups, but each individually motivated to pursue its self-interest in hiding from danger among its fellows.

If we now return to our consideration of human menstrual synchronization with these general principles in mind we might at first suppose that we have explained it, and that menstrual synchronization is a vestige of reproductive synchronization of the kind found among wildebeest. Although this explanation has been advanced, it must be said that it seems a very weak one. In the first place, it is a question of coordination, not of *birth* as in the wildebeest case, but of *ovulation*. This would be no problem if human beings, like wildebeest, were notably seasonal as far as mating was concerned because, clearly, some kind of coordinated 'mating season' would be necessary first if there was later to be an annual, coordinated season for giving birth.

Again, it would have to explain why coordination with the cycles of existing synchronized members seems to be the case, rather than cyclic coordination with some independent and presumably annual rhythm. Finally, seasonal mating would also tend to be reflected in male sexual physiology, with the size of the testes in particular varying on a seasonal basis as it does in other mammals with a mating season. Since no such effect is seen in modern men it seems strange that only female physiology should have retained evidence of the allegedly seasonal variation.

Admittedly, the human menstrual cycle does have one seemingly vague seasonal characteristic. The fact that the period which runs from one full moon to the next is twenty-nine and a half days, whereas the menstrual cycle is usually taken as twenty-eight days, has tempted some to conclude that the one must be timed to coincide roughly with the other, even though there is no evidence to suggest that women synchronize their cycles, not merely with one another, but with particular phases of the moon. The problem with this theory is that the discrepancy in the lengths of the two cycles – one and a half days – would mean that, even if lunar and menstrual months were originally in phase, they would gradually get out of step, so that it would take twenty menstrual cycles, or about one and a half years, for a period of menstruation or estrus lasting a few days to overlap once again with the same phase of the moon.

But this could be the very reason for the length of the menstrual cycle. Let us return to our central argument and look at it from the point of view of any one harem-holding male. We might suppose that, cheated of an opportunity to know when a female was most likely to conceive by her concealment of estrus, he might instead have an incentive to calculate the likely period, for instance, by observation of menstruation, which cannot be concealed. Yet any chance mutation which allowed a male reliably to predict when any particular female was likely to ovulate would require some kind of natural clock or calendar to keep account of the time that had elapsed since the end of the last cycle. Here the moon would have been ideal – if only the menstrual and lunar months had been identical in length or if one had been divisible into the other in simple whole-number ratios. For instance, if the menstrual month were sixty or fifteen days in length, the phases of the moon would have provided easily observed cues as to when ovulation was imminent. Any female who had a menstrual cycle of this easy-to-predict length would find that her estrus was no longer as cryptic as she might have wished and natural selection might then have favoured females whose cycles were so adjusted to those of the moon that the two seldom coincided. A difference of one and a half days in a total cycle length of about thirty seems an ideal arrangement if a woman's fertile period only lasts a few days. Any male who used the phases of the moon to compute a female's next estrus would be one and a half days out after only one month, three days out after two, and hopelessly wrong after three or more. Yet the near identity in length of the two cycles would mean that future coincidences of the two were so rare and widely spaced as to make the lunar calendar a useless aid to the prediction of periodic estrus. Paradoxically, then, lunar and menstrual months may be related, but not in the simple, quasi-seasonal manner that some may have supposed. Furthermore, if the theory of their relationship advanced here is correct, it shows once again that a concern with individual cost, benefit and reproductive success yields unexpected and surprising insights which are seldom granted to bland generalizations about groups and what is good for them.

Like cryptic estrus, menstrual synchronization in cohabiting females is not a characteristic unique to our species, and so need not necessarily be the outcome of hunting. It has been reported among hamadryas baboons, where it may well be a manifestation of reproductive competition among the normally unrelated females who make up a male's harem.[25] But however it came about in the first place, menstrual synchronization would have served to plug the one remaining gap in the vulnerability of hominid females to

[25]H. Kummer, *The Social Organization of Hamadryas Baboons*, pp. 176–7.

selective provisioning with meat by males who, even though they could not tell when a female was ovulating, could easily tell when she was menstruating. This is partly because human menstruation is particularly copious and unmistakable, presumably because the benefit of providing a richly vasculated lining to the womb for potential fetal implantation outweighs the cost in 'advertising' menstruation when it occurs.

Because the cost of concealing menstruation probably exceeds its benefit to any particular female, females who coordinated their menstruation to coincide with that of the other females in the group need not be at a disadvantage, since they would all be in the same position. Since the same logic applies to all nubile females in a polygynous group, all might be expected to evolve it, and evidently have done so.

With all his females who are cycling likely to be menstruating at the same time, a male would be denied the opportunity to make strategic decisions about which female to provision with meat in exchange for copulations at the expense of the others. Presumably he would either deny them all or provide equally, because all would be in the same condition. As far as any individual cycling female was concerned, neither outcome would put her at any competitive disadvantage in relation to the other females in the group. In this respect menstrual synchronization among hominid females would represent a special case of what is probably its more general evolutionary basis: the advantages which synchronized estrus can bring to females who are more reproductively successful than others by way of denying copulations wanted by other, less successful females at the same time.

But what of females who were not cycling? In all probability there would be a number of these, and any one particular female would certainly find that there were quite long periods during which her menstrual cycles were interrupted, either by pregnancy, or by lactation (which, as we have seen, inhibits ovulation if suckling occurs as frequently as it does among primal hunter–gatherers). Since a woman conceals ovulation anyway, its absence during lactation is no kind of cue that a male could use. Furthermore, the absence of menstruation would presumably be a benefit because a lactating, non-ovulating female is available for copulation throughout the month and so perhaps might be provisioned throughout it.

This is especially important because as weaning gradually occurs ovulation will begin again and it is presumably in a male's interest to re-inseminate his females promptly so that they can provide more offspring as soon as possible. And of course, it would not pay a male to avoid an obviously pregnant female if the offspring she was carrying was his own.

In summary, synchronized menstruation can be seen to fit into the same adaptive picture as cryptic estrus and both to correspond to what is, essentially, a situation of *reciprocal altruism*: one in which two unrelated parties exchange benefits to their mutual advantage. It seems possible that male hunters were able to trade meat for copulations, and females, copulations for meat, in a reciprocal relationship which has existed right down to modern times institutionalized as male 'bread-winner' and female 'wife and mother' stereotypes.

However, theoretical studies of such reciprocal relationships stress deception as especially predictable, and, if we regard cryptic estrus as a deception, we can see that here, as elsewhere, such expectations appear to be fully justified.[26] Menstrual synchronization, by contrast, seems to be less of a deception and rather more of a spoiling tactic, one which aims to equalize a disadvantage not so much with regard to the other party in the exchange (the male), but in relation to competing players on the same side (other females).

Paedomorphosis

If we return to a consideration of male adaptations for a moment, we will recall that, as I pointed out earlier, the first hunters were likely to have been young, unmated males. They would have constituted the members of the all-male groups which inevitably must have existed side by side with the one-male, polygynous groups suggested by our species' similarities to the gelada baboon, our degree of sexual dimorphism and the relative size of human testes. If we accept the possibility that females may have exchanged matings for meat it follows that those matings would have been exchanged, initially at least, with younger, unmated male members of all-male groups.

[26] Badcock, *The Problem of Altruism*, essay 1, 'Reciprocal Altruism, Repression and Regression'.

Immediately we see that the principle which seems to apply to sexual targeting in modern polygynous hunter–gatherer societies – that it might pay young men to target older women, and vice versa – might equally have applied in the distant past. This is especially so if we concede that older females, perhaps less jealously guarded by their harem 'owners' and more assertive anyway, might have had more opportunity and motivation to solicit extra-mural copulations with other males. But however that may be, the overall effect must be that female choice for meat meant female taste for matings with the younger males who could have been the only ones who could provide it.

If we now recall the general principles of sexual evolution which we reviewed earlier, we will immediately notice an intriguing possibility. This is that male evolution could have been deeply disturbed and redirected by such an effect, especially if we recall the general principle that female choice crucially affects the evolution of males, thanks to relatively high female sexual discrimination.

The precedent for this can be found in the foregoing analysis because, as we have just seen, males with meat to exchange for matings may have exercised some degree of choice with regard to the apparent sexual status of females. If those who originally favoured only estrus-advertising females left the greater number of offspring, they could have thereby conditioned the loss of an estrus signal and the development of permanently prominent breasts in modern women. With this in mind, we can now go on to ask what kinds of corresponding effects female choice may have had on the evolution of males.

In effect, a female taste for meat and matings with younger, hunting males has the effect of increasing the relative reproductive success of such males *while they are young*. This is in complete contrast to the norm for a polygynous species where some considerable measure of maturity is usually a prerequisite for male reproductive success, whether through inter-male conflict or female choice. In the first case, that of conflict among males, maturity usually means larger size and the complete development of whatever weaponry or defensive paraphernalia evolution has conferred on the males in question. In the case of female choice, a male's appeal to females will usually take time to acquire and more completely developed decorations and attributes will tend to be more desirable to females than incompletely finished ones, so that maturity will once again be an essential attribute of male reproductive success.

But the situation which we are envisaging is completely different and, in my view, goes some considerable way towards explaining the unique characteristics and unparalleled standing of our species in nature. It predicts that females may have begun to favour matings with immature males – males who were not fully developed in either behaviour or appearance. In other words, the beginnings of hunting may have had the effect, via female choice, of reversing any tendency towards increasing sexual dimorphism and may have conferred an adaptive advantage on males who matured either less or more slowly than others. Not only would this have reversed what was presumably a long-term trend towards sexual dimorphism in our species, it would also explain both the behavioural tendencies discussed earlier, and the anomalous growth of the brain which sustains them.

In fact there is evidence that such a process of retarded maturation, called *neoteny* (literally, 'staying young') results in apparent *paedomorphosis* ('taking on the form of a child') and that this has indeed been a major factor in human evolution. According to Stephen Jay Gould – an authority who, it should be noted, cannot be suspected of any Freudian or sociobiological proclivities – 'Neoteny has been a (probably *the*) major determinant of human evolution.'[27]

As we have already noted, there is no doubt that modern human beings and modern apes share a common ancestor:

> Humans and chimps are almost identical in structural genes, yet differ markedly in form and behavior. This paradox can be resolved by invoking a small genetic difference with profound effects – alterations in the regulatory system that slow down the general rate of development in humans.[28]

Again,

> adult man possesses numerous, detailed morphological characteristics which are entirely due to minor phylogenetic changes in the *rate* of growth and development in the corresponding bodily parts, and are

[27] S. J. Gould, *Ontogeny and Phylogeny*, p. 9.
[28] Ibid.

not caused by a deviation from the general *direction* of developmental differentiation common to at least all higher primates.[29]

To put the matter another way we could say that, if compared with chimpanzees, modern human beings bear a number of obvious similarities, but that they resemble the immature or even unborn forms of those near relatives much more than they do the mature, adult forms. Like the fetal chimpanzee, the adult human being is hairless; like the adult human (especially the human male), the fetal chimpanzee's first hair appears on the head and around the chin. Like it, modern human adults have rather generalized hands, without the particular adaptations which chimpanzee hands show at a later stage. Despite undoubted resemblances to gelada dentition, human adult teeth are more like the milk teeth of an ape than its adult teeth, and the *labia* which characterize the external female genitals of our species only appear as a transitory, immature stage in the development of female chimpanzees. As in all mammalian fetuses, the uro-genital tract of woman points forwards, rather than being parallel with the spine as it is in all adult mammals, and, like the fetal ape or monkey, mature man lacks a penis-bone (probably because it is one of the last to be ossified in the developing primate).

Above all, the human head, like the head of the fetal ape, is globular, thin-walled and characterized by relatively late-closing sutures. It is nicely balanced on the spinal column thanks to a flat, vertical plane to the face and high forehead. It contains a brain which is much larger in comparison with the body than that of the adult ape and which, furthermore, retains much of its generalized, immature proportions, not to mention its presumed capacity for learning, play and experimentation.

If hominid females in the distant past showed a preference for mating with younger male hominid hunters, they would also have been choosing these very characteristics. Consequently, it is possible that males who remained immature for slightly longer than others might have had an evolutionary advantage and could have left slightly more offspring than those who did not. Since these characteristics would probably tend to be heritable, their offspring would have tended to pass them on preferentially to their offspring, and

[29] Schultz quoted by Gould, *Ontogeny*, p. 375.

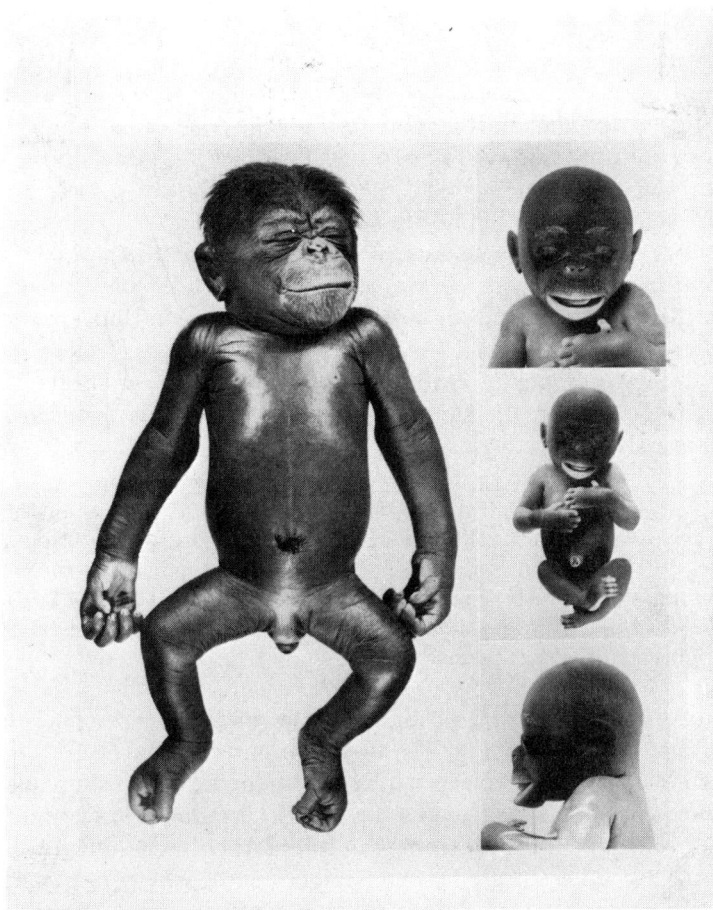

Chimpanzee foetuses at 7 months and 5 months (right)
Photograph from Adolph H. Schultz, *The Life of Primates* (Weidenfeld and Nicholson 1969, plate 13) is reproduced by kind permission of the Anthropologisches Institut der Universität Zurich, A. H. Schultz-Stiftung (photograph: Weidenfeld & Nicholson Archives)

so on in a self-perpetuating cycle of naturally selected adaptation. Furthermore, being younger in appearance, or staying younger-looking for longer and thereby having slightly greater reproductive success than others, would have had far-reaching effects on human evolution.

One such effect which would have followed directly from a female taste for meat. This is because *a more paedomorphic appearance would also have been a less sexually dimorphic one.* To understand why this is so we need to note one or two points about sexual differentiation and how it comes about.

In the early stages of embryonic development human males and females look much the same. As Darwin noted, it is generally true of mammals and birds that

> when the male differs from the female, the young of both sexes almost always resemble each other, and in a large majority of cases resemble the adult female. In both classes [i.e., birds and mammals] the male assumes the characters proper to his sex shortly before the age for reproduction; if emasculated he either never acquires such characters or subsequently loses them.[30]

The process of sexual ontogeny – the means by which each sex comes to be the way it is – is now comparatively well understood and fully bears out Darwin's insight. In the case of mammals like ourselves, it seems that males develop by gradually differentiating from an initial condition from which females diverge much less. An obvious example would be the male's testes, which begin in the same location as that in which the ovaries always remain, but migrate to the scrotum just before or after birth. Again, pathology relating to sex hormones like testosterone both before and after birth strongly suggests that males become male thanks in large part to hormonal stimulation of tissues which, if not so stimulated, would remain more female than male in appearance and function.

Sexual dimorphism is usually a function of inter-male conflict if it is not one of female choice, and presumably a reduction in sexual dimorphism (in this instance, *because of* female choice) would also tend to reduce conflict by means of de-emphasizing provocative dimorphic signals and secondary sexual displays. Indeed, we have already encountered examples of this in species with non-

[30] Darwin, *The Descent of Man*, p. 297.

dimorphic, paedomorphic or transvestite males like the bluegill sunfish. Closer to our primate home, it has recently been suggested that, in the case of orangutans,

> not all males develop large body size and other secondary sexual characteristics as adults. Small body size may benefit some males by reducing the aggression they receive from larger males. Although fully grown males are almost invariably aggressive toward each other, they usually tolerate subadult males. Small body size may permit subordinate males to travel through areas controlled by larger resident males and to gain access to females without competing directly with larger, fully grown males.[31]

The significance of this precedent lies in my earlier observation that early hominid hunting or scavenging would have had to be a cooperative affair. If males are to cooperate on hunts they must inhibit or at least reduce their natural antagonism over the question of sexual competitiveness. This desirable tendency might be served, in part at least, by a reduction in apparent sexual dimorphism, normally so provocative to other males. More paedomorphic males who were more desirable to females would also presumably have tended to cooperate better with other, similar males if their reduced dimorphism enhanced, even slightly, their ability to cooperate without conflict during a hunt. If such males also resembled females more, and if in such a polygynous situation males usually have considerable dominance over females within breeding groups, then more feminine-looking males may have been more acceptable to other males than would otherwise normally be the case.

This would most certainly have been the case in relations between young male hunters and older, dominant male polygynists who, originally at least, must have controlled all the females. Whereas the latter might regard fully adult, dimorphic males as a threat to their sexual hegemony much as dominant male gelada and hamadryas baboons do, they may well have felt less provoked by younger-looking, less dimorphic males. Perhaps such males would have been more acceptable as 'followers' or 'satellites'; and perhaps this is the means by which younger, hunting hominids in fact had access to breeding females. If toleration by older, harem-holding males was the crucial factor, reduction in sexual dimorphism and apparent

[31] P. Rodman and J. Mitani, 'Orangutans: Sexual Dimorphism in a Solitary Species'.

sexual maturity may have been the key to enhanced reproductive success for young, male hominid hunters.

If reduction in dimorphism also affected behaviour, making males act less antagonistically towards others, then the effect would be enhanced to the benefit of the reproductive success of the males concerned, since their capacity to obtain meat and to sire offspring would be promoted by an avoidance of conflict, both with males of their own age, and with older, harem-holding ones. By means of an astonishing evolutionary paradox, being less male and more feminine in appearance and behaviour would have resulted in being more successful *as a male*. The factors which would explain the paradox, of course, would be the beginnings of hunting by young male hominids, the willingness of females to exchange matings for meat and the tolerance of older, dominant males of such events, thereby promoting the reproductive success of young, hunting males and their otherwise immature sexual attributes.

Although this theory of human evolution may seem strange at first sight, it is by no means without precedent in nature. I have been able to find at least one other mammalian species where a very similar trend seems to have occurred, albeit for reasons exactly contrary to those which I am suggesting in the human case.

Whereas I am proposing that our hominid ancestors were obliged to hunt in groups if any success and safety was to be assured, sheep congregate in groups to escape, not to carry out, predation. Despite – or because of – being highly aggressive animals, it has been suggested that sheep show signs of paedomorphosis to the extent that

> the female and yearling male so greatly resemble each other that it is often difficult to separate these classes. Both may be of much the same size, carry horns of similar length and thickness, and have white bellies and ample white fur on the rear margins of the legs. One can consider the female as frozen in her development at the stage of a young ram, while the ram continues to grow and develop toward his ultimate, mature body form at 8 or 9 years of age. Male and female sheep pass through the same developmental stages, but whereas the female's development is stopped at sexual maturity, the male's development goes on.[32]

Thanks to this close resemblance of yearling males to females, 'one can claim that the young male acts like an estrous female, mimicking her behaviour and appearance, which allows him to live side by side with larger males.'

[32] Geist, *Mountain Sheep*, p. 131.

The male groups are homosexual societies in which the dominant acts the role of the courting male and the subordinate the role of the estrous female. The dominant male treats all sheep smaller than he is, irrespective of age and sex, like females; it is his prerogative to act sexually.[33]

These observations of wild sheep suggest a human analogy. Hunting by groups of young male hominids and matings with females of older harem-holders may have been the evolutionary setting in which human homosexual behaviour in particular and many male sexual 'aberrations' in general found their first and perhaps their greatest adaptive value. It may well be that the theory of human cryptic sexuality set out above finds its actual origins here: both in the conditions which affected males in hunting groups and in those which consequently made themselves felt on the females whom they provisioned. Behaviours which seem aberrant and maladaptive today in the context of modern human societies may have had a very different significance in the primeval conditions which we are considering.

Here may lie the origins of another sexual peculiarity of our species which is so pervasive and 'obvious' that we fail to see just how exceptional it is. What I have in mind is *covert copulation* – the fact that human beings universally (and this is true of primal, Australian hunter–gatherers) avoid open sexual intercourse and prefer to copulate out of sight of others. Such reclusiveness is not common among mammals in general or primates in particular. On the contrary, in most polygynous species, harem-holding males mate openly with their females, unafraid of witnesses. Not so human beings, where covert, secret copulations seem to be yet another aspect of our species' general tendency to cryptic, concealed sexuality. It is one which may go right back to primal conditions where males had to mate in secret because of the risk of interference by other, possibly older males who controlled the females.

Above all, passive, masochistic and submissive behaviour towards other males may have had a powerful adaptive value in terms of the inevitable conflict between young hunters and older, established harem-holders whose females the former could secretly seduce with meat. Perhaps such dominant males could themselves be seduced by the same means. But however that may have been, one cannot

[33] Ibid., pp. 132 and 131.

help thinking that submissiveness to such paternal figures, along with other pseudo-feminine behaviours, may have very much eased and defused the confrontation. Where young males had to pacify older, paternal ones in order to promote their own reproductive success, homosexual behaviour becomes adaptively comprehensible and immediately intelligible from an evolutionary point of view.

Again, a clear parallel is seen in the behaviour of sheep:

> Mounting is performed by dominant sheep on subordinates irrespective of the latters' sex and age. It is not only the sexual pattern during which copulation occurs but also a pattern which is the privilege of the dominant to perform. Only if a sheep can mount another without being punished has it demonstrated dominance ... Not only do dominant rams treat subordinates like females, but subordinate males may also act like females by showing lordosis [the characteristic mating posture of ewes] when mounted or urinating to the larger rams.[34]

To return to the human case, it seems that both the prototype and the crucial setting for homosexual behaviour in both sexes may be found in the two unisexual groups which the beginnings of hunting brought into contact with one another: the all-male groups of young, otherwise unmated hunters, and the harem groups of females presided over by older, dominant males. It may well be that all modern homosexual behaviour in either sex reaches back to these primeval social structures and should be considered as having been adaptive for them, rather than to anything else. Finally, when we consider the enormous simplification and unity which this brings to our understanding of the sexual aberrations as a whole, we cannot ignore the possibility that this is a promising way of looking at things.

'This is all very well,' my critic replies, 'but there is one fact which will just not fit your theory. If we look back to the figure on page 57 we can clearly see that, even though, as you yourself pointed out, the testes of modern men are small by comparison to those of chimpanzees, the penis itself is far larger. How can this be reconciled with reduced sexual differentiation? A large penis is hardly a paedomorphic characteristic. On the contrary, it bears all the appearance of being a provocative sexual signal comparable to phallic displays in other primates!'

[34] Ibid., p. 139.

I admit that it is indeed true that modern men are equipped with a penis which is both relatively and absolutely larger than that of any other primate, so that whereas the average erect human penis is about thirteen centimetres in length, that of the chimpanzee is about eight centimetres and that of the gorilla a mere three centimetres![35] But consider the following observation: the sexually dimorphic characteristics which I am assuming were lost because of paedomorphosis were presumably permanent characteristics like pronounced brow-ridges, canine teeth, and a generally 'robust' physical form. What is notable about the human penis is that it is both conspicuous when flaccid and can become unusually large for a primate penis *in erection*.[36]

Because primal hunter–gatherers, like more modern Australian aborigines, went about naked, the penis of the male would be on permanent display. Erection, however, is an intermittent phenomenon. It would be an obvious – but only temporary – indication of sexual excitement or interest. As an intermittent sexual display it might be designed specifically for display to females, rather than to provoke males in the way in which a permanent dimorphic feature might. In other words, I would tend to agree that it is indeed a phallic signal comparable, for instance, to the highly coloured penises of other male primates, but I would make the qualification that it is one intended to avoid unnecessary provocation of other males.

Only the provocative, male-directed, sexually dimorphic characters would be selected against by the mechanism of paedomorphosis; so penis enlargement, admittedly not a paedomorphic characteristic, would evolve independently, perhaps even as some kind of compensation for the loss of the other characteristics. After all, secondary sexual characteristics, sexual displays and sexual dimorphism in general do not relate just to inter-male conflict, but also to female choice. If my theory puts the latter in the forefront, a primary, but intermittent sexual display directed preferentially at females might easily be retained or even enhanced as part of the overall pattern of what was essentially a process of sexual selection

[35] R. V. Short, 'Sexual Selection and Its Component Parts, Somatic and Genital Selection, as Illustrated by Man and the Great Apes', pp. 138, 145 and 150.

[36] R. V. Short, 'Sexual Selection and the Descent of Man', in J. H. Calaby and C. H. Tyndale-Biscoe (eds), *Reproduction and Evolution*, p. 15.

relating to features judged desirable in males by females. Here, presumably, sexual quality as evidenced by the penis played some part. A female who was impressed by any particular male in this respect might expect that her male offspring by him would inherit a similar appeal to females in the future – with all that implied for her own eventual reproductive success.

To revert to my earlier discussion of the 'psychological penis', we can see that this term seems to apply with some force, not merely to the phallic period of childhood, but to the genital phase of maturity as well. Again, the development of pubic hair, albeit present in both sexes, suggests a decorative function and one, furthermore, contradicting the general paedomorphic tendency towards the loss of body hair. But the fact that in English law, for example, indecent exposure is a crime which can only be committed by a man against a woman suggests that the penis does indeed carry a psychological significance not attaching to the genitals of the female. It appears to be one which is regarded as intrinsically provocative and perhaps especially directed towards females, rather than other males.

Returning to our central theme, but putting the matter another way, one might say that paedomorphosis would have the effect of *infantilizing sexual maturity* (because it made sexually mature individuals look less mature) and of *sexualizing infancy* (because it made younger-looking individuals sexually potent[37]). Infantilization of sexuality or sexualization of infancy is essentially what Freud found to be the foundation of the Oedipus complex in particular and of infantile sexuality in general.

In this respect paedomorphosis can be seen to have had the effect of sexualizing an immature form. It would be an exact analogy of the precocious salmon parr mentioned earlier – in other words, one in which an immature, non-dimorphic form became precociously sexually mature. The difference would be that, where precocious salmon parr and bluegill sunfish 'sneakers' represent alternative types of sexually mature males, human males correspond to only one physical type showing reduced dimorphism and notable paedomorphosis.

In the human case it seems that paedomorphosis has infantilized *behaviour* just as much as it has done the physical form of males so

[37] Something which has sometimes been called *progenesis* (Gould, *Ontogeny*, p. 227). For my part, I doubt the value of such fine distinctions and prefer to use *paedomorphosis* as a term covering both neoteny and progenesis.

that much human sexuality has come to suggest 'play' and to incorporate infantile elements derived from childhood. Freud's classical oral, anal and phallic periods are, in this sense, evidence both of sexualized infancy and infantilized sexuality in adults – paedomorphosis in sexual behaviour. Activities which are otherwise primarily associated with adults' relationships with children, such as cuddling, kissing, tickling and hand-holding, reappear in the repertoire of adult sexual behaviour as unmistakable paedomorphic elements. Terms like 'foreplay' and phrases such as 'playing around', 'playing with oneself', 'on the game', 'fun and games', and many others of a similar kind attest to the fundamentally paedomorphic nature of much adult sexual behaviour. Finally, fantasy, the purely imaginative, mental equivalent of childhood play and the pleasure principle, achieves an extraordinarily important role in adult sexual life which psychoanalysis has done more than anything else to reveal.

In other words, as paedomorphosis infantilized adult sexuality, it sexualized infancy and made infantile Oedipal behaviour a possibility. This came about in two ways. Hunting created conditions where increased parental investment in offspring was a possibility. Not only did it provide a new, highly nutritious and easily portable food resource which could be carried back to home bases to provision mothers and young, it also made it possible for the young to exploit the newly available resources by lengthening childhood and vastly increasing the amount of parental investment which could be made in them.

Immediately we notice that this is reminiscent of an idea which we encountered earlier: the paradoxical value of regression to the individual offspring in its competition for investment from its parents. Paedomorphosis – adopting a more 'regressed', child-like appearance than one would otherwise have had – seems to be a kind of anatomical regression in which the offspring not merely behaves in a less mature manner, but actually takes longer to mature:

> Some recent studies indicate that retardation begins early in human development and increases continually throughout embryogenesis ... the appearance of 147 stage marks in the prenatal development of mouse and human ... is essentially the same in both species, but early stages take two to four times as long to develop in humans while later stages take five to fifteen times as long.[38]

[38] Gould, *Ontogeny*, p. 366.

In other words, paedomorphosis has not merely resulted in adult human beings taking on a more 'regressed', child-like form, it has resulted in childhood and immaturity itself becoming vastly extended. Consequently, whereas apes like the chimpanzee, gorilla and orangutan take eleven years to grow to full maturity, human beings take twenty; whereas female chimpanzees reach sexual maturity at about nine years of age and gorillas at about seven, humans average round about thirteen. Furthermore, in so far as features like degree of hair covering, skull morphology, and penis-bone development are concerned, human beings essentially *never* mature.[39]

Although not directly attributable to paedomorphosis in the way in which many other adaptations evidently are, *menopause* is presumably an inevitable consequence. This is because the lengthening of childhood and immaturity directly implied by paedomorphosis exposes a woman's children to the prospect of being orphaned long before they are mature. This is especially likely if, as is indeed the case, female mortality is strongly linked both to risks involved in pregnancy and childbirth and to increasing age. Having reached a point some ten years or so from the end of her average life-expectancy, it might pay a woman to avoid the risk of death involved in any further pregnancy and instead concentrate on seeing her existing children through to some kind of maturity and independence. In this way menopause might evolve, not as a directly paedomorphic trait, but as one which was indirectly determined by the overall trend in that direction.

In more general terms, it seems as if developmental retardation is related both to extending the period of parental investment and to the final, paedomorphic adult form of modern human beings. Indeed, we can now see that these are different sides of the same coin: adults are paedomorphic because of retardation of childhood development; childhood development is retarded because paedomorphosis became a valuable adult adaptation.

The question which seemed so paradoxical to us when we examined it before – that of explaining regression, not merely in childhood, but in adult life – now finds a more complete and unexpected answer. Now we can see that regression occurs in adult life because modern human adulthood is a fundamentally regressed condition: one in which immature appearance and behaviour have taken on a

[39] Ibid., p. 368.

major adaptive value. But whereas physical and behavioural regression in childhood is explained in terms of its value in soliciting enhanced parental investment and therefore relates to *parental* effort, the comparable regression in adulthood appears to relate mainly to reproductive success and thereby to *mating* effort.

The key to understanding how the latter could have occurred lies in realizing that mating success came to imply success in acquiring meat, whether by hunting it cooperatively in the case of males, or soliciting it in exchange for matings in the case of females. Once females had acquired it, juveniles could also acquire the increased parental investment which it represented by the Oedipal and regressive means suggested earlier.

In other words, hunting created new conditions which favoured less mature males in terms of mating success and more retarded offspring in terms of possibilities for the reproductive success of both those males and the females who chose to be impregnated by them. Such female choices in turn put a premium on males retaining the appearance of being young, resulting in a reduction in sexual dimorphism and an extension of childhood and immaturity. This in turn provided the setting for behavioural paedomorphosis, with infancy becoming sexualized because of the investment-soliciting value of Oedipal behaviour, and also infantilizing mature sexual behaviour because of the fateful part which infantile sexuality came to play in determining the sex roles of adults.

Indeed, given that paedomorphic appearance was now a factor in sexual selection, paedomorphic sexual behaviour may now have been positively selected, so that mothers might promote their own eventual reproductive success by responding, not merely to infantile sexual cues in sons (and here enlargement of the penis in childhood may have played a role analogous to that which I suggested above), but by favouring the very evidence of infantilized sexual behaviour itself as a new and effective sexual strategy.

As anatomical sexual dimorphism was reduced by paedomorphosis, so might behavioural dimorphism have become attenuated with the result that infantilized sexuality in general might have been rewarded, and not merely 'masculine', 'polygynous' behaviour. What Freud called the polymorphous perverse tendencies of childhood were probably positively favoured in this way, to some extent dissolving sharp divergences between masculine and feminine behaviour by way of Darwin's observation, noted earlier, that paedomorphosis and feminization are often much the same. As we have already seen but can now perhaps appreciate with much more clarity, this may eventually have allowed for the appearance, not merely of paedomorphic, infantilized sexual behaviour, but cryptic sexuality, with some chromosomal males behaving like females, and some genetic females behaving like males.

If the theory advanced above regarding the evolutionary origins of Oedipal behaviour and the modern paedomorphic form of human beings is correct, we can begin to see that male transvestism in particular and homosexuality in both sexes in general is in no way surprising in these circumstances, and that these factors constitute little more than carrying to an extreme a progressive reduction in sexual dimorphism which is otherwise completely normal for our species. From the purely behavioural point of view, it would represent an exaggeration of infantilized, non-dimorphic behaviour to the point where the sexes were identified to the point of confusion.

Women, anorexia and the evolution of the brain

'But,' my imaginary critic replies, 'what about women? So far you have only explained why men have evolved in this way. How can you explain the paedomorphic appearance of women? This is especially significant because, as you know, human females seem to be somewhat more fetalized in respect of hairlessness, body-build, cranial structure, and so on. You pointed out that the conventional theory linking hunting with human evolution had problems explaining the evolution of women, but it seems that your theory is just as defective because, for it too, hunting only seems to affect males.'

To these criticisms I would make the following replies. First, I would point out that it is entirely predictable that modern women would appear to be more paedomorphic because, as Darwin noted, a more paedomorphic form usually is a more feminine one, thanks to the fact that the female of the species does not acquire the sexually distinctive appearance which an adult male does.

Secondly, I would have to point out that, in so far as my argument applies to paedomorphosis reducing signs of sexual maturity in general, it does indeed apply to women and that we have already discussed what is perhaps the most importance instance of this: cryptic estrus. Since immature primate females do not advertise estrus even if they will do so later during maturity, a failure on the part of adult human females to do so is evidence of a paedomorphic tendency in their sexual behaviour.

Furthermore, this is especially convincing if we recall two earlier observations. The first is the fact that the external genitals of modern women resemble those of young female chimpanzees much more than they do the spectacularly estrus-advertising genitalia of adults. The second is the observation that, in so far as breast-development might be seen to be an estrus signal which has become permanent, it has a clear parallel in the gelada baboon, where, as we saw earlier, something analogous occurs during the estrus periods of *young* females.

Finally, I would also point out that the retardation of maturity which is responsible for the enormous prolongation of human infancy and childhood appears to be an adaptive ploy which benefits offspring of both sexes. After all, if little boys can delay maturity so as to solicit enhanced parental investment, so too can little girls. If such retardation of the process of maturity pays both sexes in childhood it is not surprising that both should exhibit its effects in adulthood.

Thus, both as a strategy to solicit parental effort, and as a factor in the mating effort of adults, retardation of maturation seems to have affected both sexes and to have benefited both. In males, conflict with other males may have been a prime factor in promoting paedomorphosis, resulting from the appeal of meat to females and its value in acquiring matings. Females, for their part, were involved in another kind of conflict with other females over access to such provisioning which resulted in the paedomorphic suppression of visible estrus. In both sexes, therefore, paedomorphosis meant reduced conflict with members of the same sex and enhanced appeal to members of the other. Far from explaining only paedomorphosis in the case of males, the theory advanced here provides an economical explanation of its evolution in both sexes thanks to a single principle: female choice of hunting hominids.

As far as brain growth in particular is concerned, it is certainly true that hunting must have remained a predominantly masculine affair. But if the theory of human evolution via paedomorphosis is correct then the human intellect and the large brain which supports it did not evolve *so as* to facilitate hunting. On the contrary, it seems that initially, a large brain/body weight ratio was an almost 'accidental' side-effect of another adaptation: namely, paedomorphosis.

The paedomorphic theory of human evolution allows us to escape from the circularities and improbabilities of the currently accepted theory. As we noted earlier, this sees brain enlargement as evolving in order to facilitate the hunting adaptation. But now we can envisage the possibility that human brain enlargement and the expansion of the human behavioural repertoire which goes with it are a consequence essentially of sexual, rather than economic or dietary factors.

But, according to the view suggested here, an enlarged brain, along with other paedomorphic characteristics such as relative hairlessness, was part of a wider picture centred on the adaptive value to hominid females of males who were young, able to hunt and somewhat less sexually dimorphic than their seniors. The elaborated behaviour which reduction in dimorphism also implied made full use of the expanded brain capacity which went along with it to produce the observed effects: cooperative hunting by males, cryptic estrus and synchronized menstruation in females, Oedipal behaviour, cryptic sexuality, extended infancy and childhood and so on.

It seems, then, as if the riddle of behavioural regression in adult life and the riddle of hominid brain growth both have the same solution: a paedomorphic retardation of development which both increased relative brain size and widened the behavioural repertoire of our species to include purely psychological equivalents of the cryptic forms of sexuality found as physical adaptations elsewhere in nature.

If we glance at the facts related to human brain growth for a moment we can see just how true the latter assertion is. Whereas the brains of most mammals are fully formed at birth, primate brains typically continue to grow for some time, so that macaques achieve 65 per cent of cranial capacity by birth and chimpanzees 40 per cent. However, new-born human beings have only completed 23 per cent of eventual brain growth, and whereas chimpanzees and gorillas will have completed 70 per cent of it by the end of their first year, human beings only reach this proportion at the end of the third year. In general, marked immaturity of the skull as evidenced by late closure of the cranial sutures can be detected until far into adult life so that not only are the human brain and skull paedomorphic in appearance, but chronically retarded in their pace of development.[40]

The paedomorphic theory suggests that parents of both sexes utilized increasingly complex brains for increasingly complex behaviours, such as the repression of memories of their own Oedipal past and the compartmentalization and subdivision of consciousness which this involved. With a premium on countering deceptive infantile tactics such as regression and Oedipal behaviour, mothers in particular needed subtle intuition and advanced interpretive skills to meet the challenge set them by their offspring. As paedomorphosis became a self-sustaining trend in hominid evolution and our ancestors made the transition from vegetarian foraging to hunting and gathering, the evolutionary foundations for the modern Oedipus complex, both positive and negative, male and female, were laid.

The theory proposed here also suggests that the general principle set out earlier which sees apparent sexual aberrations such as homosexuality as consequences of the evolution of cryptic sexual behaviour, might also apply to some disorders which appear principally to affect women. I am thinking of the eating disorders to which I referred briefly earlier. With the foregoing in mind, it seems that we might be able to add anorexia and bulimia to a list which already includes fetishism, transvestism, paedophilia, narcissism and so on.[41]

In each of those cases we saw that the apparent perversion concerned could be explained by one simple evolutionary consideration: that the behaviour served to disguise a 'normal', heterosexual orientation by substituting another, apparently unrelated one. However, all the deceptions used had one factor in common. They all sought to minimize conflict with members of the subject's own sex by appearing either to make their subjects seem uninterested in the opposite sex, or to make them resemble the other sex (or a younger, or less dimorphic form of their own sex). Furthermore, we began to understand why such aberrations are so much more common among men than among women and found

[40] Gould, *Ontogeny*, pp. 371–3.
[41] See above, pp. 134–8.

the answer in the essential nature of masculinity: small and numerous sex cells with resultant relatively less discriminate choice of object in the vast majority of cases.

The evolutionary considerations set out in this chapter suggest that, although males may generally be more disposed to conflict with their own sex and to sexual aberration than females, one important exception exists in the case of our species. This exception is the case of female competition over male provisioning which, as we have seen, may well explain numerous characteristic adaptations of modern women such as cryptic estrus, pronounced breasts and synchronized menstruation.

If the general principle is true, that apparent sexual aberrations and the forms of psychopathology built upon them originate in situations of conflict with members of the subject's own sex, we might see anorexia–bulimia as a female case in point. We might conclude that the conflict in this particular instance could be traced back to competition among hominid females for provisioning with food by the first hunters. We have already seen that my theory suggests that the latter gave rise to the particular adaptations just mentioned, but now we can envisage the possibility that it might also have given rise to a characteristic modern disorder centred on food, body-weight and sexual competition.

In my view anorexia has exactly the same psychological function as transvestism, narcissism, paedophilia and so on. Characteristically, it first occurs at puberty or in young adulthood. Since weight-gain appears to be the determining factor in triggering menarche in human females,[42] it seems that the self-starvation characteristic of anorexia has the effect of inhibiting menstruation (a condition known as *amenorrhoea*) and, if it occurs during puberty, even of reversing physical changes associated with the onset of sexual maturity. In other words, *anorexics become more paedomorphic:* they lose the characteristic attribute of nubility – sexual cycling – and revert to a more immature, less sexually differentiated form, that of the pre-adolescent.

Once again, sexual and psychological pathology seems to amount to an attempt to reverse dimorphic changes and revert to an earlier, non-sexual state. Because women are already female there is no attendant element of feminization such as we would expect in

---

[42] Frisch, 'Fatness and Fertility', p. 70.

males, but the paedomorphic aspect seems to be undeniable. In a sense, self-starvation to the point of failing to menstruate is just an exaggeration of the trend in female evolution which was begun with the concealment of estrus, which suppressed the cyclic signal. Anorexics go one step further and suppress the sexual cycle itself.

Although psychoanalysts have long known that anorexia possessed a sexual dimension, they have had to explain it in terms of the three traditional perspectives of psychoanalytic theory. The dynamic view explained the paradox of its association with the exactly opposite behaviour – bulimia – on the basis of the sound psychological principle which says that every mental action can have (and often does have) an equal and opposite reaction. Thus whereas the latent conflict may give rise to self-starvation as a primary symptom, it can equally cause the opposite behaviour – gorging. Like depression and the mania which can alternate with it in a comparable way, these extremes of behaviour signify an inability to resolve a fundamental conflict which lies beneath both alternatives and a corresponding inability to reach a normality which lies between them. In its turn, the structural/topographical view was able to account for the repression of the sexual aspect, and the economic/quantitative one the chief reasons for its onset at puberty (which is obviously the increased libidinal demands which the EGO faces at this time).

If the foregoing account of human evolution and the crucial role which female choice played in it is credible, it seems that we can now add a fourth dimension. It is one which suggests that anorexic disorders rest on a profound trend in human evolution which dictated that female sexuality, body weight and the provision of food should be deeply interconnected. It also suggests that the origin of such disorders almost certainly lies predominantly in the sufferer's sexual conflict with members of her own sex over questions of choice of members of the opposite sex. In so far as this conflict originates in childhood – and, as we have just seen, the general paedomorphic trend in human evolution predicts that it will – the conflict in question can hardly be anything other than the female version of the Oedipus complex.

But whereas traditional psychoanalysis could only situate the disparate factors of body-weight, food, femininity and sexual conflict on the three classical dimensions, it could never really explain *why* those factors had to relate in that way. It could not explain why they seemingly did so in a typical, recurrent manner which hinted that, under the purely individual and accidental determinants in each case, more general determining causes might be operating. The fourth, evolutionary dimension of psychoanalytic theory suggests that this is indeed the case and that, alongside purely personal dynamic, quantitative and topographical factors, a more general, evolutionary, adaptive one also exists which ultimately sets the scene in which the modern disorder unfolds. It seems that in a typical way anorexics give evidence, not merely of a personal Oedipal conflict developed in the course of their individual histories, but of an adaptation based on primeval competition for reproductive success which becomes pathological only in the extent, not in the nature, of its development.

If this were indeed so, then modern anorexics could be seen as victims of an evolutionary process whose predominant effects had otherwise been felt by males. Well may some modern women criticize the shortcomings of modern men, but, if the line of reasoning being pursued here is correct, it is clear that to a very considerable extent modern man is the creation of woman and that the other side of the coin of male sexual indiscriminateness is female discrimination which shapes the evolution of males. If, in the polygynous mating system typical of our species, nearly all females but only some males mate, and if females have some considerable element of choice about which males do mate with them, it follows that females will exercise a privilege of control over certain aspects of male development almost completely denied to males in respect of females. In other words, if the connections between female choice, hunting and human evolution are indeed as I have proposed in this essay, then anorexics give expression to one of the most surprising possibilities raised by modern views of natural selection: the realization that modern man may be less the creature of God than he is the creation of primeval woman.

Conclusion:

Psychoanalysis in Evolution

Evolution beyond the pleasure principle

The three essays which make up the bulk of this book have concentrated directly on the subject of sexuality and, by way of a necessary psychological consequence, have indirectly concentrated on the workings of the pleasure principle. The reasons for this were set out in the introduction. In conclusion I would like to round out and complete the picture by alluding – albeit perhaps too briefly – to the large and important part of human psychology and behaviour which lies, in the title of one of Freud's most famous books, *Beyond the Pleasure Principle*.

Within its pages he elaborates on *the compulsion to repeat* as a factor in the unconscious more fundamental still than the requirement that pleasure be maximized and pain avoided. Essentially, he saw this as the basis of one of the strangest findings of psychoanalysis, what he came to call *transference*, and which can be understood as an unconscious compulsion to model new experiences and relationships on previous, repressed and forgotten ones.

Without this effect, psychoanalysis could hardly be possible, as Freud himself clearly saw. This is because it was the compulsion to cast the analyst in the role of the protagonists in the patient's neurosis which both made the latent conflict visible in the analysis, and provided a means of resolving it. At first through exploiting the suggestive power of the analytic transference (usually to the analyst as a parent), but later through the full analysis of the so-called *transference neurosis* (the compulsive recreation of the conflict in the analysis itself), Freud found that a conscious, EGO-mediated resolution was possible – at least as an ideal of analytic therapy.[1]

The reappearance of the latent conflict in the contemporary conditions of the analysis could hardly be seen as motivated by the pleasure principle since its chief effects were seldom if ever minimization of pain. On the contrary, since repressions existed to safeguard against the anguish caused by the repressed, the undoing of repression in analysis usually produced greater conscious conflict and more emotional disturbance than had existed previously, at least temporarily. In this sense the transference neurosis was a new symptom, a secondary, transitory disorder attendant on the primary one.

Admittedly, its appearance was ultimately motivated by a desire to recover from the conflict and to maximize pleasure in the sense of minimizing neurotic misery. Indeed, the ideal analytic cure should restore the capacity for instinctual gratification usually so severely compromised by the illness. But in itself transference, both within the analysis and outside it, was a product, not of the pleasure principle, but of the compulsion to repeat. For this compulsion, Freud could offer no real explanation, except to say that it was in the nature of instincts to wish to restore some previous state, and since the previous state of living matter had been inorganic matter, life instincts were based on an instinct which strove for death.

Although this is fine metaphysics, it is doubtful psychology and very poor biology. As an example of how a new, evolutionary dimension might transform such problems as this one, let me suggest the following, alternative explanation.

If we consider transference in its widest context as a compulsive tendency to remodel existing experiences on earlier ones, I believe it can become entirely explicable as a classical Darwinian adaptation in the conditions of hominid hunting and gathering societies. In such conditions, individuals spend most of their time in small groups, most of whose members are usually kin. If we make the reasonable assumption that human behaviour is in large part heritable and that similar environmental conditions will have similar effects on closely related individuals, it is quite possible that transference behaviour might evolve as a kind of unconscious, compulsive awareness of precedent.

[1] For a fuller discussion of this point see Badcock, *Essential Freud*, pp. 106–10.

For instance, individuals with whom one interacted early in life would tend to have descendants with whom one might interact later in life, possibly in similar circumstances to the original ones. If those descendants inherited part of the determinants of their behaviour from their predecessors, an unconscious expectation of some similarities in one's own interactions with them would be justified.

This would be all the more so if consciousness and verbal thought had evolved to be what they evidently are – namely, relatively superficial and highly vulnerable to social influence, plausible persuasion and temporary factors, passing whims and insubstantial logic. Individuals may claim to be different and to be free of the vices or weaknesses of their predecessors; but the unconscious might render one a service by refusing to believe it. Bad experiences in particular could leave deep psychological scars in the form of negative or ambivalent transferences which can turn out to be fully justified in the grim light of experience. In such circumstances expecting the worst may be safer than credulously believing the best.

An example might be provided by the following kind of case which, although hypothetical, must have actually occurred innumerable times in human evolution. Primal hunters and gatherers, like modern ones, are dependent on highly unreliable food resources. Game can vanish or not be where it is normally to be expected and plant crops can fail or give meagre returns. Flood, drought and the general vagaries of the weather can lead to unpredictable foraging and hunting conditions. Starvation can and does sometimes occur.

Memory of such unpleasant events tends to be suppressed by the pleasure principle, which motivates all such repression. Intended as it is to direct the EGO towards rewarding, productive and successful responses to the demands of the ID, it tends to avoid anything suggestive of pain, conflict and failure. Yet certain experiences, such as starvation brought on by drought, could be worth remembering if similar conditions were to recur again. Here the more short-term, temporary considerations which influence the pleasure principle might come into conflict with the longer-term interests of the individual which could indeed be favoured by a memory of the earlier event, no matter how unpleasant. This is where a compulsive, latent expectation, albeit of a very undesirable outcome, would serve to prepare and motivate the individual to survive in circumstances which the pleasure principle might not wish to envisage, but which transference, operating at a deeper level, nevertheless would not forget.

If this line of reasoning is correct it suggests that a compulsive awareness of precedent – what we otherwise would call 'transference' – could evolve both counter to the pleasure principle and in accordance with it. In the latter instance the pleasure principle and transference conspire to produce the characteristic *fixations* of the libido. These would constitute precedents whose compulsive character derived from much more than pleasant memories of the past and whose notorious intransigence would consequently be explained as a result of reinforcement of the pleasure principle by a mechanism powerful enough to overcome it and, in this case, fully justifying the term Freud coined to describe it: the compulsion to repeat. In *Beyond the Pleasure Principle* he traced this compulsion to the nature of instincts in general, but a consideration of it in the light of the theory advanced here could explain one of the most puzzling aspects of the compulsion to repeat: the so-called 'traumatic neuroses'.

Not surprisingly, a consideration of these recurs a number of times in *Beyond the Pleasure Principle*, along with the concession that the wish-fulfilment theory of dreams appears to break down when confronted with the fact of recurrent dreams of traumatic events which can hardly have been wish-fulfilling. The fact that individuals who have suffered traumatic experiences in war, vehicle accidents or other situations frequently suffer from the repetition of the trauma in dreams or hallucinations argues strongly for some kind of compulsion to repeat and clearly indicates the limitations of the pleasure principle, whose preferences they so clearly flout. Yet the theory advanced here could explain them, especially if we regard them as a modern effect of an adaptation originally acquired in circumstances where traumatic frights, although by no means impossible, might have been a lot less likely.

We are certainly on safe ground if we argue that war as we know it does not occur in primal hunter–gatherer economies, and what might pass for it comprises desultory skirmishing between small groups who always have the option of getting well away from trouble, should they need to do so. Very low population density plus a nomadic way of life means that large concentrations of manpower for purposes of war are simply not feasible. Furthermore, in an economy without storable wealth and with necessarily fluid geographical and social boundaries, the occasion for such wholesale conflict hardly ever exists; never to the extent which was to become common after the Neolithic Revolution.

Admittedly, unexpected, traumatic shocks could still occur. Ambushes and sudden raids could happen. But in such conditions individuals would normally have some kind of warning or expectation, and sharp-eyed hunters who spend their lives stalking prey are not likely to be surprised very often. Furthermore, when such a shock does occur, it will come in the form of enemies armed with weapons for hand-to-hand fighting and for hunting. It will most emphatically not come in the form of sudden explosions, gun-fire out of nowhere, aerial bombing or the unexpected detonation of mines.

Again, primal hunter–gatherers, reliant on their feet to get them about, are hardly likely to experience violent collisions or crashes; and, since they live in the open, fire and mayhem in an enclosed space will be a similarly very rare or non-existent experience. Admittedly, earthquakes or volcanic explosions can occur without warning and be totally traumatic; but the chances of such events occurring in any one lifetime must be very low. Furthermore, although sudden flash floods and storm damage are also possible, lightning never strikes out of a blue sky, and some kind of preliminary warning is usually available if weather conditions are about to turn catastrophically bad.

Finally, in conditions in which illness, disease and death are much more common than in advanced industrial societies and in which average life-expectancy is significantly shorter, the kinds of emotional traumas incident on such things are likely to be lessened somewhat by their very frequency, so that death and loss, although always likely to be distressing, can hardly be unusual.

In short, it is possible that the traumatic neuroses which so puzzled Freud and which constituted such a severe test of the pleasure principle need to be seen in their evolutionary perspective as the result of traumas unlikely to afflict primal human beings to anything like the extent to which they affect modern men and women. Their mystifying recurrence in the symptoms of the traumatic neurosis or in dreams would be explicable as the workings of the mechanism of transference in the context of a traumatically painful experience whose impact had been so overwhelming that compulsive memory of it almost seemed to be as great a shock as the original experience itself. Here the trauma would be a reflection of modern conditions, the compulsion to repeat it an adaptation acquired long before those conditions came into existence.

In modern psychoanalysis the concept of transference has, like so much else, been 'medicalized' and restricted in its meaning. Many regard it as applicable only to the *analytic transference*, where the analyst is compulsively identified with protagonists in the patient's past and thereby becomes ineluctably a part of the transference neurosis. Yet it seems to me arbitrary and unjustified to restrict the term in this way because transference as a phenomenon is not restricted to the analytic situation. On the contrary, it is funda-mental to the psychological process which produces groups and has manifold effects in other respects. Almost inevitably, those who undergo an analysis will find that the analyst takes on identifications and transference roles previously occupied by others whom the analysand has encountered. To call this 'transference' when the subject of the identification is an analyst but something else when it is not seems to me unjustified and arbitrary; and while I can readily see that the restriction of this term to analysis might do something to enhance the mystique of the analyst and the analytic process, where scientific considerations predominate I believe it has no justification.

The usage suggested here returns it to something of its original meaning, where it was allied to the concept of *displacement*, for instance, in dreams, the context in which Freud first described it. Displacement means the linking of a particular, repressed, latent content with a quite different manifest content, which serves to give it a 'disguised' representation, so to speak. Under the influence of displacement latent contents masquerade as something else and thereby escape recognition.[2]

By linking the enlarged concept of transference with Freud's 'compulsion to repeat' and tracing both back to primal conditions, it is possible to establish a more basic mechanism, founded not in the nature of instincts as such, but in a redefined ID, understood as a set of fitness-maximizing demands placed upon the EGO. Among these are transference demands: compulsions either in accordance with or, more crucially, contrary to, the pleasure principle which

---

[2] Freud, *The Interpretation of Dreams*, V, 562; 'Analysis of a Phobia in a Five-Year-Old Boy', X, 51.

require the EGO to act in accordance with precedents which may not be registered in consciously accessible memory but which have been retained in the unconscious. Contemporary events or relationships which arouse these latent memories act like displacements of them: they link new manifest conditions to previous, latent ones which thereby become activated as a compulsive force of precedent within the EGO which deserves to be seen as transference in its widest context.

Death and the depressive adaptation

The mention of death and bereavement just now suggests one final example of how limitations on the pleasure principle might come about without recourse to death instincts as such. The very existence of mourning behaviour and the depression which is usually a part of it will provide one further and very telling instance of how a fourth, evolutionary perspective can go way beyond the pleasure principle to the basic adaptive demands which that principle itself first evolved to fulfil.

It is significant that, whereas Freud refers to the traumatic neuroses and their recurrent dreams in *Beyond the Pleasure Principle*, there is no mention of 'mourning and melancholia' (to quote the title of another of his works), despite a lengthy discussion of the life and death instincts. In the paper which does bear that title, mourning and what we would today call depression rather than melancholia are discussed mainly in quantitative terms.

This is entirely justifiable in view of the obvious fact that what Freud called *the work of mourning* – the laboriously slow detachment of the libido from the mental representations of the lost object – appears to require a definite amount of time to be carried out. This, in turn, suggests a definite amount of libido being detached at a definite rate, dependent on the amount of time spent by the EGO in its characteristic preoccupation with the loss. Again, the restriction in the EGO which characteristically accompanies mourning, graphically suggested by Freud's description of 'the shadow of the object' falling across it, is entirely comprehensible in terms of the reproaches which the individual feels against the lost object for becoming lost but which, through identification with it, are instead directed at the self. Such internalized aggression is subjectively perceived as depression and guilt; the diminution of the EGO as sadness and melancholy; the disappearance of the object as emptiness, frustration and loss.

Admittedly, in *Beyond the Pleasure Principle* Freud does relate sadomasochism to the death instinct, but in analysing the dynamics of mourning and its attendant depression he could content himself with a mainly quantitative and topographical view, rather than the qualitative one implicitly introduced with the dualism of the life/ death instincts. Such an approach seems most unpromising in view of our almost total lack of knowledge about the constituent substances involved in psychological economics, and so here, once again, I would like to suggest an evolutionary, adaptive alternative.

In the case of mourning in particular and depression in general there can be no question of any very apparent gratification of the pleasure principle. One wonders, then, why these behaviours evolved in the first place. If we look at mourning as a kind of 'natural', 'normal' incidence of depression we might begin to find an answer.

On the face of it, we might expect the pleasure principle to motivate people *not* to feel any loss or unpleasure when someone near to them died or disappeared. After all, why cause oneself needless distress? Nothing can bring back the dead. Yet, for all that, human beings consistently behave *as if the dead might come back*. They grieve for them, want them to return, seem to be constantly reminded of them, almost seem to expect them to return to the point where others can be momentarily mistaken for them. Those who remain alive often experience the 'guilt of the survivor'; they wish that the deceased had not died and compulsively level recriminations at themselves in case there was any way in which their death could have been prevented.

But why should this be so? Why is it not adaptive immediately and unemotionally to accept that the dead are dead and can never return? Why waste such mental and physical effort in mourning the dead when one might so easily just forget them? After all, according to both psychoanalysis and sociobiology, human beings are especially adept at forgetting whatever is in their interest not to remember.

The short and simple answer to this question is that, in our primal hunter–gatherer past, those thought to be dead easily might return. Look at it this way: primal hunter-gatherers wander in small bands over vast areas on their endless quest for food. Only comparatively seldom will someone die in the sight of all their near relatives and friends, and those who are too ill to move on with the group after local food resources are consumed will usually be left to their fate. In circumstances such as these it is likely that death will be represented as it always seems to be in the unconscious: as a disappearance, but as a disappearance which always implies the possibility of a return. This aspect of it was graphically revealed in an anecdote told by Freud. It was of a ten-year-old boy whose father had died and who is reported on one occasion to have remarked to his mother, 'I know father's dead, but what I can't understand is why he doesn't come home to supper?'[3]

Analytic experience teaches that in the unconscious there is no such thing as death, only disappearance and that, as far as the unconscious is concerned, disappearance always connotes the possibility of return. The reason for this is straightforward: in primal hunter–gatherer economies death usually is a disappearance – those who die are those who do not come back or who, if left somewhere, will never be found again. But this fact also introduces an element of uncertainty in that return is by no means impossible, although necessarily its likelihood declines with time.

According to this theory, the possibility of return explains the quantitative element so apparent to Freud and to anyone who has ever had to undertake 'the work of mourning' themselves. Because disappearance cannot be taken to mean death in all cases it is not necessarily in a person's long-term interests instantly to act as if the temporarily absent are the finally dead. If they were known to be gone definitely for ever, redistributing their property and re-arranging the social network of the group without them would be justified and usually in someone's – occasionally in everyone's – interests. But a sudden reappearance after premature reorganization could cost the individuals concerned a great deal.

A man who, like the widow's mite, was lost but then found, would take it very badly if, after an absence of only a short while, other men had taken his wives and children, his weapons and hunting rights, and would feel very much abused if his position in the social network had been usurped, his friends and allies become those of others and his whole standing and identity null and void. Faced with such extortionate costs, the benefit of conflict to regain all

---

[3] Freud, *The Interpretation of Dreams*, IV, 254n.

that he had lost would be potentially enormous and the cost to others with whom he had to contest in that instance correspondingly great. Comparable observations would apply to women.

In short, given the vagaries of a nomadic existence, it would not be in any individual's own self-interest necessarily to conclude after only a short disappearance that someone was gone for ever. Instead, it would pay them to wait and see. But how should they wait, and for how long should they see?

As far as waiting was concerned, it would really be a question of not performing those acts with regard to the lost individual and their property which the pleasure principle might otherwise suggest. For instance, if there had been an emotional or sexual tie with the lost person it would not pay to liquidate it instantly and reassign the relationship to others. In quantitative terms it would pay to take some time over reallocating libido invested in the lost individual and so a process of laborious detachment like that described by Freud in connection with mourning would be expected.

Again, a heavy cost might be incurred by premature reshaping of the social network around a missing person, and so a tendency to expect their return and to linger over realignment would be in the long-term interests of all concerned, given that only time could tell whether the absence in question was temporary or permanent. Here again, a slow process of gradual acceptance of the loss would be expected, along with real sensations of bereavement and sadness about the lost individual. Ambivalence above all would have to be safeguarded against, and here the positive aspects of one's feelings about the missing person would have to be exaggerated to act as a defence against the negative, hostile ones.

The net effect of this would be the turning back against the self of reproachful feelings described by Freud as characteristic of mourning. But this would only be the dynamic expression of an evolutionary adaptation: one which promoted the inclusive fitness of the individuals who possessed it, not by any simple or obvious means, but by the subtle, unconscious logic involved in assessing costs and benefits of behaviour directed towards important objects whose return could not immediately be ruled out of account.

In general, the attendant depression and melancholy of those who mourn would reduce their aggressiveness and initiative with regard to the opportunities presented by the loss. In time, these opportunities could perhaps be judiciously realized, but immediate, pleasure-maximizing action in their regard could easily be very costly. Therefore the adaptive advantages of a period of waiting might tend to outweigh the benefits of immediate, callous and unreflective action to get what one could out of the situation. Finally, the appearance of having grievously felt the loss of the missing person would powerfully recommend one to them were they to reappear after all, and damage done by separation could soon by mended by the observation of genuine feelings of joy at being reunited. Perhaps even the insomnia which usually accompanies bereavement might pay if its consequence was to show one awake and expectant when the lost one returned.

But what of those who did indeed die in the sight of their associates? Surely no such adaptation of delayed takeover of their position or possessions would be relevant in their case? And what of those whose position was always marginal and of minor importance, or whose assets and relationships had already been usurped? Surely, in all these circumstances mourning would be unnecessary and counter-productive?

These are certainly real difficulties, but I do not think that they are insurmountable. In the first place, it is important to realize that, where evolutionary adaptations are concerned, the norm counts for more than the exception, and unusual circumstances inevitably will make a behaviour evolved for a typical situation seem out of place in an untypical one. It seems likely that the majority of individuals would indeed die out of sight of primal hunter–gatherer groups and that mourning as an adaptation may be suited to this typical situation.

If this were so, even those who did die in plain sight of their kith and kin might be mourned because, in the unconscious, death retained the connotation of loss rather than final extinction – a supposition fully justified by modern psychoanalytic investigations. Furthermore, if ambivalence is usually involved – and, here again, psychoanalytic findings urge that it is – such contradictory motives and feelings will be mobilized whether the fact of death is certain or not. The use of the positive, loving side of the ambivalence to safeguard against the negative, hating side will still tend to emerge once death has done the ultimate harm to someone that anything can and will still tend to leave the positive feelings towards them in temporary ascendancy. Inevitably, gratification of hatred by the death of the object will free whatever more pleasant feelings exist towards it so that these will tend towards the same result: mourning of the dead, with ambivalence about the loss tending to inhibit immediate initiative in abolishing their role.

As far as the marginal and unimportant are concerned, it seems that since mourning is directly proportional to an individual's psychological involvement with the lost object, those who have little significance for others are likely to be mourned less extensively anyway. In primal societies young children might certainly fall into this category, and the elaborate funeral arrangements made for rich and powerful people in advanced, post-Neolithic societies certainly suggests that mourning behaviour is directly proportional to the psychological and social importance of the deceased.

Finally, the fact that mourning behaviour seems to be an adaptation especially evolved in human beings suggests that something about the primal hunter–gatherer adaptation may have made it especially important. The elaborate greeting and recognition ceremonies found in other mammalian social hunting species suggests that it is the cooperative hunting adaptation which makes psychological ties within the group especially important. If death or disappearance breaks such ties, we must expect corresponding ceremonial, and human mourning rites seem to have evolved to meet such a need, along with the mourning behaviour on which they are based.

If we now ask for how long the period of mourning should last, the answer would appear to be: at least a year, but not much more than two. This has the advantage of being in accordance with our knowledge of the length of normal mourning and agrees nicely with the theory proposed here. This is because, even though primal hunter–gatherers may have inhabited a tropical habitat, seasonal factors would still have been apparent in relation to flowering and fruiting seasons of various plants and migrations of game (for instance, as dictated by annual 'dry' and 'wet' seasons). The migrations of hunters are usually closely coordinated with such seasonal factors and it follows that someone who disappears at a certain time of the year might, if they were to reappear, do so at much the same time a year later or in much the same place as was visited in the previous annual migration of the hunters. Failing this, others who might have seen or heard of the lost individual might be encountered at these times, leading to an expectation that the anniversary of the disappearance might be especially important in judging the likelihood of their return.

This observation explains the painful importance of anniversaries in mourning.[4] Under modern conditions the calendar takes over the function previously performed by the cycle of the seasons in reminding the living of the lost and priming the former with special expectations of finding the latter when the same time of the year came round. The parallel importance of visiting a place associated with those presumed dead at such a time could also have its origin in the fact that the wanderings of primal hunter–gatherers frequently take them back to the same places they had visited exactly one year before.

However, after two such cycles had passed, even nomadic hunter–gatherers would probably be safe in concluding that the lost individual was gone for ever, or, at least, was unlikely to reappear in the immediate future. At this point reallocation of their kin, possessions and position to others would appear as fully justified as it was ever likely to be. Mourning would cease and the libido involved in it would be free. Only at this point would the pleasure principle reassert itself and the mourning adaptation be suspended.

The recent identification of Seasonal Affective Disorder (SAD) as a clinical diagnosis suggests that the link between seasonal changes and depression can become very real, at least in this case. Furthermore, there is reason to suspect that, if mourning behaviour is a Darwinian adaptation of the kind suggested here, pathology could affect either half of it: the tendency to depression and self-recrimination, or the timing mechanism. As far as the latter is concerned, it could easily become an independent pathology related to the cycle of the seasons, as it evidently does in the aptly termed SAD syndrome, or it could malfunction in other ways.[5]

An analogy is suggested by an alarm clock I once had. Following an accident, it acquired the habit of giving a short ring every hour, so that, when set to go off shall we say at 7.30 am, it would reserve its full response for that time, but give a minor ring every hour, on the half hour. Presumably much the same could occur with the timing mechanism involved in the mourning adaptation, so that something originally evolved to work on an annual cycle, perhaps cued by the seasons, could start to malfunction on the basis of some other cycle, such as a daily rhythm. Conceivably, this could result in

[4] G. H. Pollock, 'Anniversary Reactions, Trauma and Mourning'.
[5] R. J. and J. J. Wurtman, 'Carbohydrates and Depression'.

the oscillations occasionally found in so-called *cyclothymia*, which constitutes violent swings in mood on a recurring pattern, often regularly timed and sometimes on a 24-hour basis.

Turning now to the other aspect of mourning behaviour, the depression and remorse to which it also gives rise, it seems that this too can become detached from its obvious adaptive role and function independently. Unlike cyclothymic behaviour, this need not necessarily be pathological because it seems that depression might serve adaptive functions independent of mourning. Although it is undoubtedly a complex phenomenon, its essence seems to lie in the same fundamental mechanism as that which I suggested with regard to mourning.

Animals who undergo a change for the worse in their social status can sometimes manifest notable hormonal changes; in the case of males usually the result of a reduction in testosterone levels, making them less aggressive and sexually active. Presumably this is adaptive, in the sense that it reduces the likelihood of a male 'wasting' effort in fruitless dominance-maintaining behaviour after his dominant position has been lost. It certainly underlines the very high cost of dominance to the dominant individual. For instance, male gelada baboons who are displaced as harem-holders experience a sudden ageing which is not infrequently followed by a prompt death.[6] This effect, sometimes suggestively called 'hormonal castration', might correspond to the psychological depression of aggressive behaviour which usually accompanies the onset of depressed behaviour in men.

Nevertheless, a clear parallel to psychological depression and remorse is found in the self-directed aggressive behaviour of some animals who will, for instance, bite themselves if, being at the bottom of the dominance hierarchy or otherwise unable to find a victim, they have to take their aggression out on themselves. Perhaps this is defensive in the sense that it signals to other, dominant animals that they need not punish it further; and perhaps human depression functions in a similar kind of way. Perhaps, just as depression of individuals' libidos and aggression can be adaptive in the context of the disappearance of an individual with whom they have intimate social relations, it can also be more generally adaptive in saving those who depress their own behaviour from becoming

[6] Dunbar, *Reproductive Decisions*, p. 132.

the victims of suppression at the hands of others. In this respect it would correspond to a kind of psychological purdah, a veiling of the EGO in depressive affect which, like actual veiling of women in some modern Moslem countries, would serve as a necessary safeguard against physical assault by the religious police in the event of not effacing oneself in this way.

According to psychoanalytic insights, suicide, apparently the most supremely non-adaptive behaviour, is usually a question of the victims killing themselves rather than some other. If this is so, then suicide provides the most extreme case of the general principle that depressive and self-damaging behaviour (among which we must include guilt and remorse, self-deprecation and self-recrimination) is indeed an adaptation, despite its apparent conflict with self-interest. Its contradiction of the pleasure principle certainly suggests that there is indeed something beyond it which, as Freud intuited, has deep biological roots. But rather than being rooted in a death instinct as such, remorse, depression and the whole of adult human response to death may reflect a more complex, but more credible reality: a set of fitness-maximizing behavioural adaptations acquired in the million-odd years of human hunter–gatherer prehistory.

If this line of reasoning is correct, then mourning behaviour in particular and depression in general must be seen as belonging, not to the ego as has tended to be maintained, but to our new, redefined ID. This is not to deny the value of Freud's brilliant dynamic description of mourning, nor the fact that in its course, 'the shadow of the object falls across the ego.' It most certainly does; my point is that the shadow is cast by a light switched on in the unconscious in response to a classical Darwinian behavioural adaptation: mourning behaviour which evolved as a kind of genetically determined brake on the pleasure principle and the egoism of the individual in the face of temptations aroused by the absence of the object.

Here, a clear paradigm for depression in general would be supplied by post-natal depression. As I have suggested elsewhere,[7] there are reasons for thinking that a period of post-natal ambivalence about a baby might pay a mother faced with many years of parental investment in it if a few days of relative neglect served to test the baby's health and ability to survive. The Mexico

---

[7] Badcock, *The Problem of Altruism*, pp. 16–17.

City earthquake of a few years ago showed that new-born babies – in sharp contrast to their mothers – could survive up to two weeks' burial in the rubble of a maternity hospital and still be brought out alive.

In large part, this is thanks to the unusually large amount of subcutaneous fat with which human babies are born, but that factor in itself may well be a co-evolved adaptation to post-natal depression in human mothers. While the pleasure principle might tend to motivate a mother to accept her baby readily and derive gratification from it, something 'beyond the pleasure principle' might want to put a brake on her immediate acceptance in the interests of her longer-term reproductive success. In short, post-natal depression might evolve to test a baby and save its mother from futile investment in an infant which might not have the health and strength to reach adult life and pass on their mother's genes.

The ego-analytic view of mourning behaviour as defensive and the evolutionary view followed here are complementary, but one or two differences of emphasis and insight can be detected. In the first place, the evolutionary description of mourning as an innately determined behaviour suggests that it cannot and should not be prevented or regarded in any way as 'pathological' or even undesirable. Unpleasant it will always be, but attempts to minimize or curtail it for that reason are likely to do more harm than good, and modern attitudes to death and mourning – not to mention post-natal depression – may not be entirely helpful in this respect.

In the second place, we must notice an important difference with regard to depression. Psychoanalysis has traditionally viewed defences from the viewpoint of the ego: they have been seen as attempts by the ego to deal with inner conflicts or external threats which could be traced to the ego's current predicament. It has not considered something like depression as an automatic, innate defence which originally evolved in primal hunting and gathering societies to deal with external threats from others.

But if the theory proposed here is correct, depression is a common modern symptom with prehistoric, evolutionary roots which make the ID demand that the EGO direct aggression against itself rather than express it openly against others whenever modern conditions create situations which might have made that depressive reaction adaptive in the past. To return to my earlier analogy of purdah, it is as if women went on veiling themselves long after the policeman with his truncheon ceased to enforce such conformity and then found it hard to understand why they should continue to wish to do so. If wearing purdah were written into the unconscious demands of the human ID it would express itself exactly as modern compulsions do: an automatic, irrational behaviour which seemed to have no justification in the present but whose meaning was locked away in the evolutionary past. Its origins would lie beyond the present and beyond the pleasure principle, not in instinct as such, but in what created such compulsions in the first place: our evolutionary, primal hunter–gatherer past.

Freud, Galileo and psychoanalytic 'canals on Mars'

One welcome effect of rethinking psychoanalysis along these lines may be a change in existing attitudes, both to psychoanalysis in general and to Freud's work in particular. As far as Freud is concerned, it is almost certain that the future will take a somewhat different view from that current until now. The history of science is full of examples of such changes, and one which seems to me to be particularly pertinent to Freud is the case of Galileo.

Like Freud today, Galileo in his own times and for some considerable time afterwards was regarded as a self-confessed revolutionary and propagandist for radical views which seemed perversely contradictory to logic, commonsense and everyday experience. A moving Earth seemed self-evidently absurd, because not only did it not seem to move, but the sun evidently did cross the sky just as surely as a bird, cloud or any other object did. If this seems naive, subtle thinkers pointed out that objects dropped fell at one's feet, not to one side as a moving Earth seemed to indicate; and had anyone been able to repeat that experiment with a ray of light in the sixteenth or seventeenth century its result would have seemed conclusive: the Earth could not move.

Again, lacking a principle of universal gravitation, a belief in the motion of a spherical Earth through space seemed violently counter to commonsense, not to mention accepted views of motion. In a similar way, the Freudian concept of the unconscious and its counter-intuitive, dynamic view of psychology seems absurd to many today who persevere in the notion that 'consciousness' and 'the mind' are one and the same thing and that avowed motives and thoughts are the only ones that matter.

Galileo, like Freud in the eyes of many today, was seen as a controversialist who published propaganda arguing a new philosophy, rather than the careful experimenter and tireless observer we now know him to have been. But this is only because his notebooks have been published and because his views have become universally accepted and indeed now seem obvious, rather than counter-intuitive (thanks in part to modern technology and views of the Earth from space). Today the Galileo we see is not merely the man who was persecuted by the Catholic Church for his controversial writings. Today we see a Galileo unknown before the publication of the notebooks: we see Galileo the founder of observational astronomy; Galileo the pioneer of modern experimental method; Galileo the mathematician and forerunner of Newton, Einstein and a whole profession of modern physicists whose theory and practice have retrospectively completely revolutionized our attitude to their great Italian predecessor.

Like contemporary views of Galileo, present attitudes to Freud see him mainly as the exponent of a controversial ideology, rather than as an exemplar of careful, scientific observation. He is thought of in the same context as Karl Marx, rather than in that of Darwin or Einstein, and his purely psychological concept of repression is erroneously confused with the meaning of the term in politics, in particular, Marxism.[8]

Yet, just as few today would read the dialogues of Galileo unless pursuing an interest in the history of science, so I suspect few will read Freud's theoretical, metapsychological works in the future. If the undertaking represented here is in any way indicative of the future, Freud's theoretical works are likely to be largely superseded and absorbed into future syntheses as those of Galileo were in the works of Newton and later physicists. But just as modern commentators marvel at Galileo's notebooks, with their intriguing records of astonishingly accurate experiment, careful observation and original speculation, so future generations, already in possession of theories more advanced than Freud's, will come to appreciate the detailed and painstaking observations, comments and insights contained in his more descriptive works.

---

[8] See Badcock, *Essential Freud*, pp. 9–21.

Just as the Jovian system seemed to raise more problems than it solved by posing the questions ultimately answered by Newton, but only partially answered and vaguely intuited by Galileo, so Freud's discovery of oral behaviour or the Oedipus complex, for instance, raises more questions than psychoanalysis to date has been able to answer. Like Galileo, Freud may have had to admit that the 'how' and the 'why' of it largely escaped him, but, as I have tried to show, the 'how' and 'why' of the Oedipus complex or oral behaviour can now be addressed, thanks to new insights derived from evolutionary biology.

But Galileo's sighting of the moons of Jupiter raised a more immediate question. This was: 'Is it really there?' Many doubted, especially professional astronomers of his own day, who, either lacking telescopes altogether or only having access to primitive instruments much given to imperfection and distortion, were prone to contradict the truth of his observations. Paradoxically – and analogously to the more recent case of Freud – it was non-astronomers in the main, even artists and intellectuals, who first took up his ideas, leaving the professional establishment at best sceptical, at worst openly hostile.

Some adopted a very sophisticated view, exploiting the long-standing tradition of using philosophical sophistry to obstruct scientific progress in a way remarkably comparable to what was to happen later with regard to Freud and psychoanalysis. Their argument went as follows: 'Galileo claims to see new "facts" through his telescope. The telescope works by magnifying light as it passes through lenses. Yet magnification is only a kind of distortion of the light. If the light is distorted, the image might be also. Therefore what Galileo sees through his telescope is at best distorted, at worst, unreal!'

Before dismissing this as pure nonsense, it might be worth remembering that it did contain more than a grain of truth. Given the primitive lenses of the time, distortions did indeed exist. For instance, all bright objects were surrounded by halos of rainbow-coloured light, what we would now call *chromatic aberration* (an effect produced by the lenses essentially acting as prisms, so that light was partially decomposed into its spectrum as it passed through them). At the time, some claimed that the halos were real objects. Characteristically, Galileo rightly thought that they were imperfections produced by the telescopic lenses and could safely be ignored. With the invention of the reflecting telescope (in large part as a solution to the problem) the matter was finally settled: colour-fringes were apparent, not real. Evidently, the neo-Platonists who refused to look through 'distorting' lenses at reality had a point.

They had an ever better point in the case of the alleged 'canals' on the planet Mars which many observers – among them reputable astronomers – reported during the nineteenth and early twentieth centuries. Not only this, some of the more imaginative among them speculated that the planet was inhabited and claimed that the 'canals' which they could see were evidence of planetary engineering works constructed by intelligent beings, probably much like ourselves. Today we know that these 'canals' were apparent, not real, and that although what appear to be ancient water courses are present on Mars, these are not the phenomena which the early Mars-watchers claimed to have seen.

Freud, like Galileo, found what he did largely thanks to inventing a new instrument. That instrument was not an apparatus, like the telescope, but a method of observation called *free association*. It enabled individuals, for the first time in history, to report uncensored trains of thought and to follow them wherever they went in a setting in which an observer, shielded from their view, could follow in practically ideal circumstances. By adopting an attitude of 'free-floating' attention complementary to the free association of the person on the couch, the observer – the psychoanalyst – attempted to interfere as little as possible and to allow uninhibited expression to associations which would normally be censored in any other circumstance. The result was an astonishing insight into the human mind every bit as significant and arresting as Galileo's observation of the Jovian system. Yet, like Galileo's, Freud's observation is disputed by many who use arguments fundamentally comparable to the neo-Platonist one about systematic 'distortion' rehearsed above.

Although that objection seems to me to be as unreasonable when applied to Freud's own use of his method as it was to Galileo's telescopic observations (which, given the limitations of his equipment, were remarkably accurate and sensibly interpreted), I cannot deny that psychoanalytic equivalents of 'canals' on Mars certainly exist.

Psychoanalysis, like astronomy, has had more than its fair share of cranks, and on occasions Freud himself was forced to distance himself from some of the more fanciful theories of his followers. An example might be Otto Rank's attempt to trace all anxiety to a birth trauma. As Freud noted, 'Rank dwells, as suits him best, now on the child's recollection of its happy intra-uterine existence, now on its recollection of the traumatic disturbance which interrupted that existence – which leaves the door wide open for arbitrary interpretation.'[9]

Part of the answer to the problem is suggested by the very context in which Freud's comment about Rank's theory quoted above is found. It comes from a work in which Freud undertook further extensive revision of his theory, just as he had done earlier in *Beyond the Pleasure Principle.* In it Freud discusses Rank's ideas because he was concerned to revise his own regarding the nature of anxiety. Originally, Freud adopted a largely quantitative view, seeing anxiety as a transformation of libido which had been denied an outlet. Rank, of course, saw anxiety as derived, not from such an instinctual transformation, but from recollections of the trauma of birth. Not only did this lead to the rather far-fetched interpretations mentioned by Freud in the quotation above (such as a fear of mice being derived from the analogy which their appearance from holes had with the birth trauma), it also made a subjective experience the origin of all subsequent feelings of anxiety, as if one original fright could explain all later ones.

As Freud was aware, this merely pushed the problem back to the beginning and made one ask why birth should be traumatic and why anxiety should be felt in connection with it. The fact that the capacity to feel anxiety while being born had logically to precede the actual experience meant that Rank's theory could not explain anxiety as such, and Freud felt that his own earlier theory could not either.

His final view, that anxiety is always a response to danger and that it is not a transformation of instinct, but an instinctive response to situations which threaten the ego with loss, pain and frustration is one which sees it in terms of an innate disposition, presumably produced by evolution, rather than a mere consequence of an environmental influence or experience, like being born. In other words, he sees it much more as an *adaptation of the EGO* than as a

[9]Freud, *Inhibitions, Symptoms and Anxiety,* XX, 136.

precipitate of early experience and this transforms his theory of anxiety from being a narrowly psychodynamic one into a broadly behavioural one, as if he had implicitly accepted the fourth, evolutionary and adaptive dimension of psychoanalysis proposed here. In his own words, 'we thus give the biological aspect of the anxiety affect its due importance by recognizing anxiety as the general reaction to situations of danger.'[10]

This is an important concession to the biological perspective because one cannot help noticing that it is an over-emphasis on accidental, environmental factors which often makes psychoanalytic interpretations look far-fetched. It seems to me that a shift in emphasis away from personal experiences like birth traumas towards innate, evolutionary adaptations like that suggested by Freud's final theory of anxiety would do much to save psychoanalysis from one of its worst vices – a tendency to over-emphasize the importance of factors bearing on individual development (psychological ontogeny) and to neglect the evolutionary viewpoint (psychological phylogeny).

At the very end of his life Freud explicitly recognized the limitations of personal psychological experience and began to emphasize the importance of phylogenetic factors:

> When we study the reactions to early traumas, we are quite often surprised to find that they are not strictly limited to what the subject himself has really experienced but diverge from it in a way which fits in much better with the model of a phylogenetic event and, in general, can only be explained by such an influence.

And, anticipating the findings set out earlier in the second of these essays, he continues:

> The behaviour of neurotic children towards their parents in the Oedipus and castration complex abounds in such reactions, which seem unjustified in the individual case and only become intelligible phylogenetically – by their connection with the experience of earlier generations.[11]

---

[10] Freud, *Inhibitions, Symptoms and Anxiety*, XX, 162.
[11] Freud, *Moses and Monotheism*, XXIII, 99.

Here, as elsewhere in psychoanalysis, the problem is compounded by the fact of *over-determination* – the observation that symptoms, thoughts and wishes often had more than one unconscious determinant. This called into being the technical practice of *over-interpretation*, understood as the linking of the same manifest content via associative chains to numerous different latent causes. But 'over-interpretation' can easily lead to *over*-interpretation understood in the colloquial sense of excessive interpretation, or theorizing which goes beyond the material on which it is purportedly based.

In my view, this is the other principal cause of the numerous psychoanalytic equivalents which can be found to the astronomical 'canals' on Mars. The method of free association can be misleading, particularly in the context of over-determination, so that analysts pursue numerous lines of causation, often in a rather confused way, never really sure which is most important, or even which are real and which not. Again, just like the 'canals' on Mars, no two observers seem able to agree on most of these things, and so divergent schools of thought emerge, often as internally chaotic as they are externally discrepant with one another. Just as astronomical fact regarding the planet Mars became heavily contaminated with science fiction earlier this century, so psychological fact seems to have become correspondingly adulterated with fiction in much of modern psychoanalysis.

One advantage which might be gained by the addition of a fourth dimension of psychoanalysis like that advocated here is that this, being an evolutionary and biological one, would not be based solely on clinical psychoanalytic observations. On the contrary, it would represent the point at which psychoanalysis and behavioural science were joined as a whole and would open up its findings to broader observational validation. In particular, the fact that the modern evolutionary view can establish something of a final cause for any particular adaptation means that it avoids the problem of over-determination, hitherto so confusing to analysts.

Because natural selection is the ultimate arbiter of evolutionary change and because we now feel some confidence in understanding its finer workings, it is possible both to complement psychoanalytic findings with wider biological ones and to advance final, evolutionary causes for their existence. For instance, we have seen that it is possible that oral behaviour as discovered by Freud is a Darwinian adaptation aimed at manipulating maternal fertility in the interests of the infant rather than in those of the mother. Again, we saw that preferential parental investment in sons may correspond to, and give the ultimate theoretical and causal explanation of, Freud's observations concerning the Oedipus complex in males and penis-envy in females. If these interpretations are correct, ultimate determinants of penis-envy and Oedipal behaviour will have been found, along with an insight into the reasons for their occurrence, fateful consequences and intricate complexities. If it is accepted that oral behaviour really has evolved to manipulate maternal fertility and that Oedipal behaviour and penis-envy really do underlie some aspects of the human reproductive strategy, it will demonstrate that some of Freud's most controversial observations resemble Galileo's sighting of the moons of Jupiter much more than they do the belief of some later astronomers that they had found civilization on Mars.

Bibliography

Abraham, K., 'The First Pre-genital Phase of the Libido', in *Selected Papers on Psychoanalysis*, London, 1948.

Alexander, R. D. and Noonan, K. M., 'Concealment of Ovulation, Parental Care and Human Social Evolution', in N. Chagnon and W. Irons, (eds), *Evolutionary Biology and Human Social Behavior*, North Scituate, 1979.

Alexander, R. D., Hoogland, S. H., Howard, R. D., Noonan, H. M. and Sherman, D. W., 'Sexual Dimorphism and Breeding Systems in Pinnipeds, Ungulates, Primates and Humans', in N. Chagnon and W. Irons (eds), *Evolutionary Biology and Human Social Behavior*, North Scituate, 1979.

Austad, S. N. , 'The Adaptable Opossum', *Scientific American*, 258, 2 (1988), pp. 54–9.

Austin, C. R. and Short, R. V., *Reproduction in Mammals*, second edition, Book 3: *Hormonal Control of Reproduction*, Cambridge, 1984.

Badcock, C., *The Problem of Altruism*, Oxford, 1986.

—— *Essential Freud*, Oxford, 1988.

Barash, D., *Sociobiology and Behavior*, second edition, London, 1982.

Benshoof, L. and Thornhill, R., 'The Evolution of Monogamy and Concealed Ovulation in Humans', *Journal of Social and Biological Structures*, 2 (1979), pp. 95–106.

Betzig, L., Borgerhoff Mulder, M. and Turke, P.,(eds), *Human Reproductive Behavior*, Cambridge, Mass., 1988.

Biller, H. 'Father Absence, Divorce and Personality Development', in M. Lamb (ed), *The Role of the Father in Child Development*, second edition, New York, 1981.

—— 'The Father and Sex Role Development in M. Lamb (ed), *The Role of the Father in Child Development*, second edition, New York, 1981.

Blaffer Hrdy, S., 'Sex-biased Parental Investment among Primates and other Mammals: A Critical Evaluation of the Trivers-Willard Hypothesis', in R. Gelles and J. Lancaster (eds), *Child Abuse and Neglect: Biosocial Dimensions*, New York, 1987.

Blurton Jones, N. G. and da Costa, E., 'A Suggested Adaptive Value of Toddler Night Waking: Delaying the Birth of the Next Sibling', *Ethology and Sociobiology*, 8 (1987), pp. 135–42.

Bowlby, J., *Attachment*, London, 1982.

Burley, N., 'The Evolution of Concealed Ovulation', *The American Naturalist*, 114, 6 (1979), pp. 835–58.

Charnov, E., 'Natural Selection and Sex Change in Pandalid Shrimps: A Test of Life History Theory', *American Naturalist*, 113 (1979).

Crews, D., 'Courtship in Unisexual Lizards: A Model for Brain Evolution', *Scientific American*, 256, 6 (1987), pp. 72–7.

Crook, J. H., 'Sexual Selection, Dimorphism and Social Organization in the Primates', in B. Campbell, (ed), *Sexual Selection and the Descent of Man*, Chicago, 1972.

—— *The Evolution of Human Consciousness*, Oxford, 1980.

Daly, M. and Wilson, M., *Sex, Evolution and Behavior*, second edition, Boston, 1983.

Darwin, C., *The Descent of Man and Selection in Relation to Sex*, London, 1871.

Denniston, R. H., 'Ambisexuality in Animals', in J. Marmor (ed), *Homosexual Behavior*, New York, 1980.

Dickemann, M., 'The Ecology of Mating Systems in Hypergynous-dowry Societies', *Social Science Information*, 18 (1979), pp. 163–95.

—— 'Female Infanticide and Reproductive Strategies in Stratified Human Societies', in N. Chagnon and W. Irons (eds), *Evolutionary Biology and Human Social Behavior*, North Scituate, 1979.

Dominey, W. J., 'Female Mimicry in Male Bluegill Sunfish: A Genetic Polymorphism?', *Nature*, 284 (1980), pp. 546–8.

Dunbar, R., *Reproductive Decisions*, Princeton, 1986.

Epstein, A. W., 'Fetishism: A Comprehensive View', *Science and Psychoanalysis*, xv (1969), pp. 81–7.

—— 'The Fetish Object: Phylogenetic Considerations', *Archives of Sexual Behaviour*, 4,3 (1975), pp. 303–8.

—— 'The Phylogenetics of Fetishism', in S. D. Wilson, (ed), *Variant Sexuality: Research and Theory*, London, 1987.

Evans, R., 'Physical and Biochemical Characteristics of Homosexual Men', *Journal of Consulting and Clinical Psychology*, 39 (1972), pp. 140–7.

Fisher, R. A., *The Genetical Theory of Natural Selection*, Oxford, 1930.

Foley, R., 'A Reconsideration of the Role of Predation on Large

Mammals in Tropical Hunter-gatherer Adaptation', *Man*, 17 (1982), pp. 393–402.

Ford, C. S., and Beach, F. A., *Patterns of Sexual Behavior*, New York, 1951.

Freud, S., *A Phylogenetic Fantasy*, Cambridge, Mass., 1987.

—— *The Interpretation of Dreams*, Standard Edition of the Complete Psychological Works of Sigmund Freud, London, 1953–74, v and vi.

—— *Three Essays on the Theory of Sexuality*, Standard Edition, vii.

—— 'Analysis of a Phobia in a Five-Year-Old Boy', Standard Edition, x.

—— *Leonardo da Vinci and a Memory of his Childhood*, Standard Edition, xi.

—— *Introductory Lectures on Psychoanalysis*, Standard Edition, xvi.

—— 'From the History of an Infantile Neurosis', Standard Edition, xvii.

—— 'Two Encyclopaedia Articles: (A) Psychoanalysis', Standard Edition, xviii.

—— 'A Case of Homosexuality in a Woman', Standard Edition, xviii

—— *Inhibitions, Symptoms and Anxiety*, Standard Edition, xx.

—— 'Female Sexuality', Standard Edition, xxi.

—— *Analysis Terminable and Interminable*, Standard Edition, xxiii.

—— *Moses and Monotheism*, Standard Edition, xxiii.

Frisch, R. E., 'Fatness and Fertility', *Scientific American*, 258, 3 (1980), pp. 70–8.

Geist, V., *Mountain Sheep: A Study in Behavior and Evolution*, Chicago, 1971.

Goodale, J. C., *Tiwi Wives*, Seattle, 1971.

Gould, S. J., *Ontogeny and Phylogeny*, Cambridge, Mass., 1977.

Grafen, A., 'How Not to Measure Inclusive Fitness', *Nature*, 298 (1982), pp. 425–6.

Graham C. A. and McGrew, W. C., 'Menstrual Synchrony in Female Undergraduates Living on a Coeducational Campus', *Psychoneuroendocrinology*, 5 (1980), pp. 245–52.

Gregor, T., *Anxious Pleasures: The Sexual Lives of an Amazonian People*, Chicago, 1985.

Gutzke, W. and Crews, D., 'Embryonic Temperature Determines Adult Sexuality in a Reptile', *Nature*, 332 (1988), pp. 832–4.

Hamilton, W. D., 'The Genetical Evolution of Social Behaviour', *Journal of Theoretical Biology*, 7 (1964), pp. 1–50.

Harcourt, A. H., Harvey, P. H., Larson, S. G. and Short, R. V., 'Testis Weight, Body Weight and Breeding System in Primates', *Nature*, 293 (1981), pp. 55–7.

Hart, C. W. M. and Pilling, A. R., *The Tiwi of North Australia*, New York, 1961.

Hartung, J., 'Polygyny and Inheritance of Wealth', *Current Anthropology*, 23 (1982).

—— 'Deceiving Down', in J. Lockard and D. Paulhus (eds), *Self-Deception: An Adaptive Mechanism*, Englewood Cliffs, NJ, 1988.

Hill, K., 'Hunting and Human Evolution', *Journal of Human Evolution*, 11 (1982), pp. 521–44.

Huxley, J. S., 'The Present Standing of the Theory of Sexual Selection', in G. R. De Beer, (ed), *Evolution*, Oxford, 1938.

Jolly, C., 'The Seed-eaters: A New Model of Hominid Differentiation Based on a Baboon Analogy', *Man* 5, I (1970).

Jones, E., *The Life and Work of Sigmund Freud*, I, New York, 1953.

Kinsey, A. C., Pomeroy, W. B. and Martin, C. E., *Sexual Behavior in the Human Female*, Philadelphia, 1953.

Kitcher, P., 'The Animal Within: Biology and Social Science', *LSE Quarterly*, 2, 4, (Winter 1988).

Kummer, K., *The Social Organization of Hamadryas Baboons*, Chicago, 1968.

Lamb, M., 'Fathers and Child Development: An Integrative Overview', in M. Lamb (ed), *The Role of the Father in Child Development*, second edition, New York, 1981.

Lévi-Strauss, C., *The Elementary Structures of Kinship*, Boston, 1969.

Li, S., and Owings, D., 'Sexual Selection in the Three-spined Stickleback', *Zeitschrift fur Tierpsychologie*, 46 (1978), pp. 359–71.

Liley, N. R., 'Ethological Isolating Mechanisms in Four Sympatric Species of Poeciliid Fishes', *Behaviour*, Supplement 13 (1966), pp. 1–197.

Lloyd, A. T., *On the Evolution of Instincts: Implications for Psychoanalysis*, unpublished MS.

Lozoff, B. et al. , 'The Mother-Newborn Relationship: Limits of Adaptability', *Journal of Pediatrics*, 91 (1977), pp. 1–12.

MacFarlane, J. A., *The Psychology of Childbirth*, Cambridge, Mass., 1977.

Martin, R. D. and May, R. M., 'Outward Signs of Breeding', *Nature*, 293 (1981), pp. 7–8.

Masters, W. and Johnson, V., *Human Sexual Response*, Boston, 1966.

McClintock, M., 'Menstrual Synchrony and Suppression', *Nature*, 229 (1971), pp. 244–5.

Møller, A. P., 'Female Choice Selects for Male Sexual Tail Ornaments in the Monogamous Swallow', *Nature*, 332 (1988), pp. 640–1.

Murdock, G. P., *Social Structure*, New York, 1949.

—— *Ethnographic Atlas*, Pittsburgh, 1967.

Nesse, R. M., 'Why Do Babies Spit Up', *ASCAP Newsletter*, 2, 7 (July 1989).

Nishida, T. and Hiraiwa-Hasegawa, M., 'Chimpanzees and Bonobos: Cooperative Relationships Among Males', in B. Smuts et al. (eds), *Primate Societies*, Chicago, 1987.

Parker, G. A., Baker, R. and Smith, V., 'The Origin and Evolution of Gamete Dimorphism and the Male-Female Phenomenon', *Journal of Theoretical Biology*, 36 (1972), pp. 529–53.

Pollock, G. H., 'Anniversary Reactions, Trauma and Mourning', *The Psychoanalytic Quarterly*, 39 (1970), pp. 347–71.

Pomeroy, W. B., 'Kinsey and the Institute', in M. S. Weinberg (ed), *Sex Research: Studies from the Kinsey Institute*, New York, 1976.

Porter, R. H., 'Kin Recognition: Functions and Mediating Mechanisms' in C. Crawford et al. (eds) *Sociobiology and Psychology*, Hillsdale, NJ, 1987.

Quadagno, D. M. et al., 'Influence of Male Social Contacts, Exercise and All-female Living Conditions on the Menstrual Cycle', *Psychoneuroendocrinology*, 6 (1981), pp. 239–44.

Ralls, K. and Brownwell, R. Jr., 'Sperm Competition in Grey Whales', *Nature*, 336 (1988), pp. 116–17.

Ramanamma, A. and Bambawale, U., 'The Mania for Sons', *Social Science and Medicine*, 14 (1980), pp. 107–10.

Rancour-Laferriere, D., *Signs of the Flesh: An Essay on the Evolution of Hominid Sexuality*, Berlin, 1985.

Robertson, D., 'Social Control of Sex Reversal in a Coral-reef Fish', *Science*, 117 (1972), pp. 1007–9.

Rodman P. and Mitani, J., 'Orangutans: Sexual Dimorphism in a Solitary Species', in B. Smuts et al. (eds), *Primate Societies*, Chicago, 1986, pp. 149–50.

Róheim, G., *The Riddle of the Sphinx*, London, 1934.

—— 'Psychoanalysis of Primitive Cultural Types', *International Journal of Psychoanalysis*, XIII (1933).

Rose, F. G. G., *Classification of Kin, Age Structure and Marriage amongst the Groote Eylandt Aborigines*, Berlin, 1960.

Ruse, M., *Homosexuality*, Oxford, 1988.

Santangelo, A., *Il Giardino dell'Eden*, Milan, 1987.

Shapiro, W., *Miwuyt Marriage*, Philadelphia, 1981.

Short, R. V., 'Sexual Selection and the Descent of Man', in J. H. Calaby and C. H. Tyndale-Biscoe (eds), *Reproduction and Evolution*, Canberra, 1977.

—— 'Sexual Selection and Its Component Parts, Somatic and Genital Selection, as Illustrated by Man and the Great Apes', *Advances in the Study of Behavior*, 9, (1979).

—— 'Oestrous and Menstrual Cycles', in C. R. Austin and R. V. Short (eds), *Reproduction in Mammals*, second edition, Book 3: *Hormonal Control of Reproduction*, Cambridge, 1984.

Siskind, J., 'Tropical Forest Hunters and the Economy of Sex', in D. Gross (ed), *Peoples and Cultures of Native South America*, New York, 1973.

Spiro, M. E., *Oedipus in the Trobriands*, Chicago, 1982.

Stoller, R. J., *Presentations of Gender*, New Haven, 1986.

Symons, D., *The Evolution of Human Sexuality*, New York, 1979.

Szalay, F., 'Hunting-scavenging Proto-hominids: A Model for Human Origins', *Man*, 10 (1975), pp. 420–92.

Thapa, S., Short, R. V. and Potts, M., 'Breast Feeding, Birth Spacing and Their Effects on Child Survival', *Nature*, 335 (1988), pp. 670–82.

Trivers, S. and Willard, D., 'Natural Selection of Parental Ability to Vary the Sex Ratio of Offspring', *Science*, 179 (1973), pp. 90–1.

Trivers, R., *Social Evolution*, Menlo Park, Ca., 1986.

—— 'Parental Investment and Sexual Selection', in B. Campbell (ed), *Sexual Selection and the Descent of Man*, Chicago, 1972.

—— 'Parent–Offspring Conflict', *American Zoologist*, 14 (1974).

Van den Berghe, E., 'Piracy as an Alternative Tactic for Males', *Nature*, 334 (1988), pp. 697–8.

Voland, E., 'Human Sex Ratio Manipulation: Historical Data from a German Parish', in J. Wind (ed), *Essays in Human Sociobiology*, I, London, 1985.

Weatherhead, P. J. and Robertson, R. J., 'Offspring Quality and the Polygyny Threshold: The "Sexy Son Hypothesis"', *American Naturalist*, 113 (1979), pp. 201–8.

Weinberg, S., *The First Three Minutes*, London, 1977.

Westermarck, E., *The History of Human Marriage*, London, 1891.

White, D. R., 'Rethinking Polygyny', *Current Anthropology*, 29 (1988).

Wilson, E. O., *Sociobiology*, Cambridge, Mass., 1975.

Wurtman, R. J. and J. J., 'Carbohydrates and Depression', *Scientific American*, 260, 1 (1989), pp. 50–57.

Index

primal phantasies, 93
primal scene, 91–3
promiscuity, 35–6
psychoanalysis
  anal-sadistic phase, 93
  on anorexia, 185–6
  and castration complex, 105–8,
    134–5
  and child's point of view, 67–9
  classical as opposed to revisionist,
    23–6
  and dreams, 13–14
  and drive theory, 12–14
  and Electra complex, 80, 110
  and evolutionary description,
    4–8, 70
  and feminism, 102–3
  on fetishism, 135–6
  on homosexuality, 112–16, 134,
    136, 139
  and id–ego–superego model of
    the mind, 8
  on identification, 111, 112
  infant's view of sex, 100
  and latency period, 118
  and new ID–EGO model, 8–13
  and Nirvana principle, 11
  on oral character, 77
  on penis-envy, 99, 101–2
  phallocentric, 103–4
  pre-Oedipal phase and penis-
    envy, 96
  and primal scene, 92–3
  and problems of interpretation,
    207–10
  on sex, 17–19, 27, 32–3
  and transference, 187–8
  Trivers's comments on, 78
  see also Freud; Oedipus complex
psychological penis, 95, 176

quantitative description, 4–5

Rajputs, 83
Rank, Otto, 207
recessive genes, 60
reciprocal altruism, 165

regression, 70–2, 177–8, 182
reproductive success, 2–3, 10–11,
  25, 42, 154
Róheim, Géza, 90

Sambia, 24
seasonal affective disorder, 199
sex
  cells, 19–20
  differences, 32, 58–60
  differences as observed by
    children, 103–4
  and Freud, 17–18
  and gender, 24
  perversions, 32–5, 72, 173–4
  ratios, 48, 51–3, 59, 81, 83, 138
  and society, 22, 47–8
  and sociobiology, 17–22
  transformation, 38
  see also parental investment
sexual bimaturism, 58
sexual dimorphism, 29, 56–8, 113,
  167, 170–1, 174–5
sexual selection, 29
sexual strategies, 26–41, 47–8,
  108–10, 132, 138
'sexy son' hypothesis, 86–94
Shapiro, Warren, 21, 121
Sharanahua, 149–50
Sharp, Keith, 117, 119
socialization, 67–9
sociobiology, 17, 133
sociology, 21–3
Spiro, Melford, 130
steatopygal, 159
Stoller, Robert, 23–4, 112, 113, 115,
  116
superego, 118
synchronization, 160–5

testis size, 62–4
*Three Essays on the Theory of
  Sexuality*, 17–18, 32, 70–1
Tiwi, 120–1, 122, 124–5, 129
topography, topographical
  description, 4–5

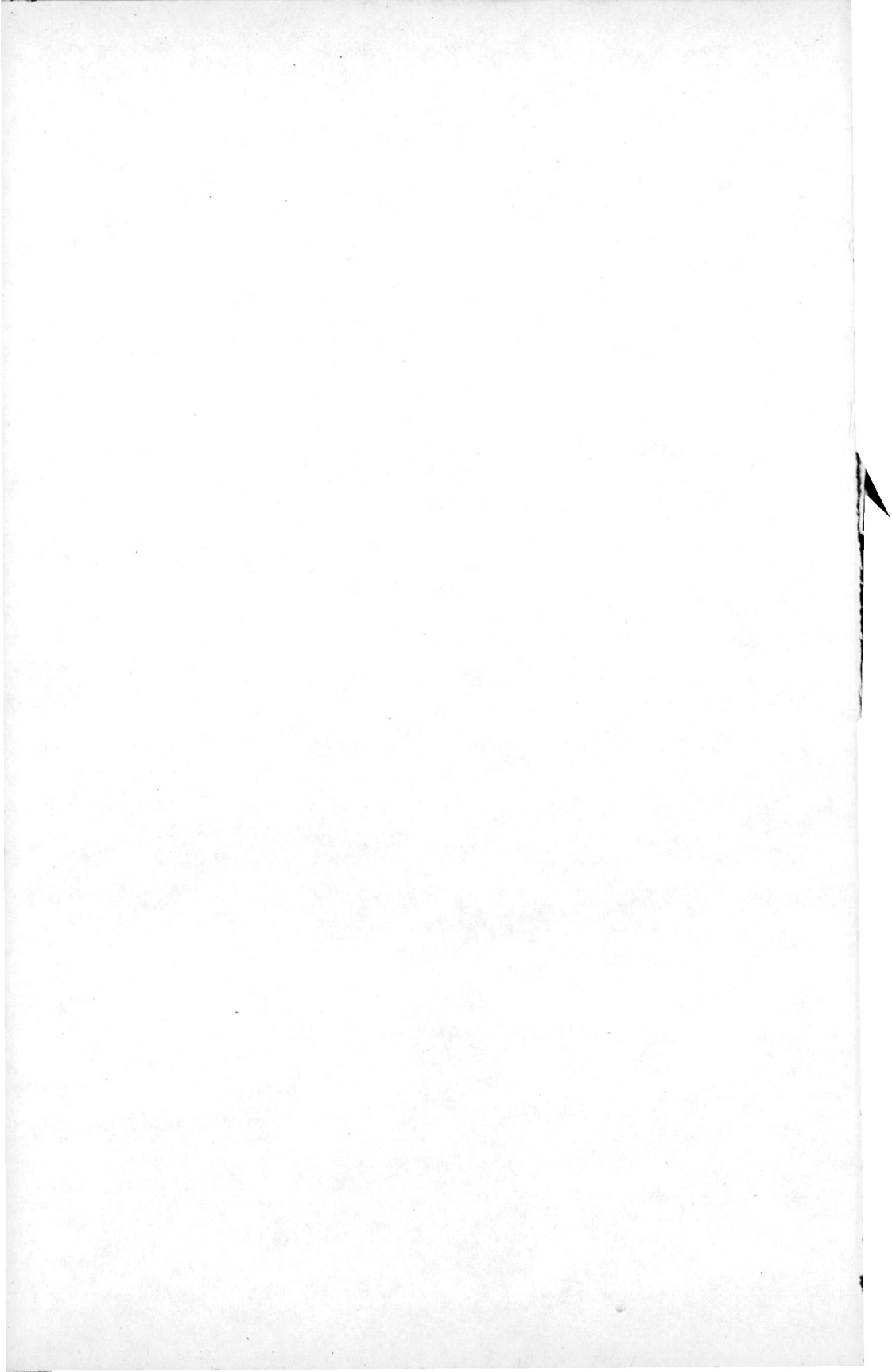